ENCYCLOPEDIA of ANIMALS

WIDE EYED EDITIONS

INTRODUCTION FROM THE AUTHOR

LOOK CLOSELY AT THE ANIMALS OF PLANET EARTH AND YOU WILL NOTICE FEATURES THAT MANY OF THEM SHARE.

There are animals with six legs, such as ants and beetles, that we call insects. There are animals with eight legs, such as spiders and mites, that we call arachnids. There are furry creatures that we call mammals. There are scaly-skinned creatures we call reptiles. There are the animals known as amphibians, many of whom temporarily make ponds a nursery ground for their tadpole babies. And then there are the dizzying numbers of fish that swim in the world's oceans, rivers and lakes. These are the great tribes of animals which make up some of what we call the 'taxonomic groups of life'.

Throughout my career in zoology, discovering the secrets of some of the animals that make up these groups has been an immense and glorious pleasure. In this book, I am privileged to share these animal secrets with you, the reader. Of course, the animals in this book are not *all* the animals in the world. At the moment, scientists have named 1.5 million different kinds of animal – yet there are many more out there for you to discover should you have the patience and skills to look, examine and understand. My hope is that this book inspires you to do just this, unlocking a host of new animal secrets along the way for others to enjoy in the future.

So, let us turn the pages and become acquainted with some of the weird and wonderful animals that call this incredible planet their home. Six animal groups; thousands of mind-boggling ways of life. Join me in celebrating some of the most fascinating.

CONTENTS

4-45	**INVERTEBRATES**	84-103	**REPTILES**
46-73	**FISH**	104-139	**BIRDS**
74-83	**AMPHIBIANS**	140-187	**MAMMALS**

HOW TO USE THIS BOOK

CHAPTER HEADER: Learn about how the species on these pages are linked together.

NAVIGATION: Use these handy tabs to work out which group of animals you are reading about.

LATIN NAME: Each species has a scientific Latin name that is the same in all languages.

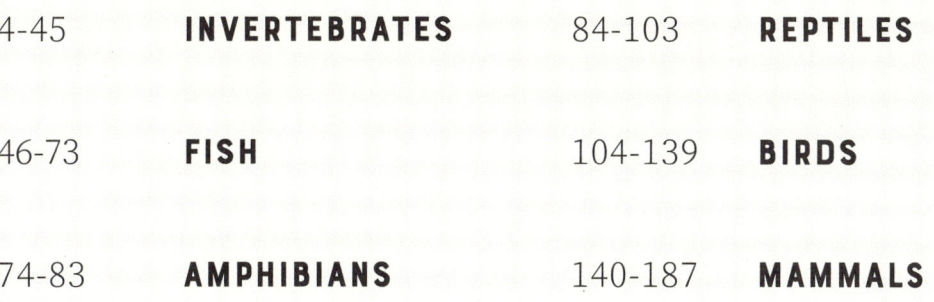

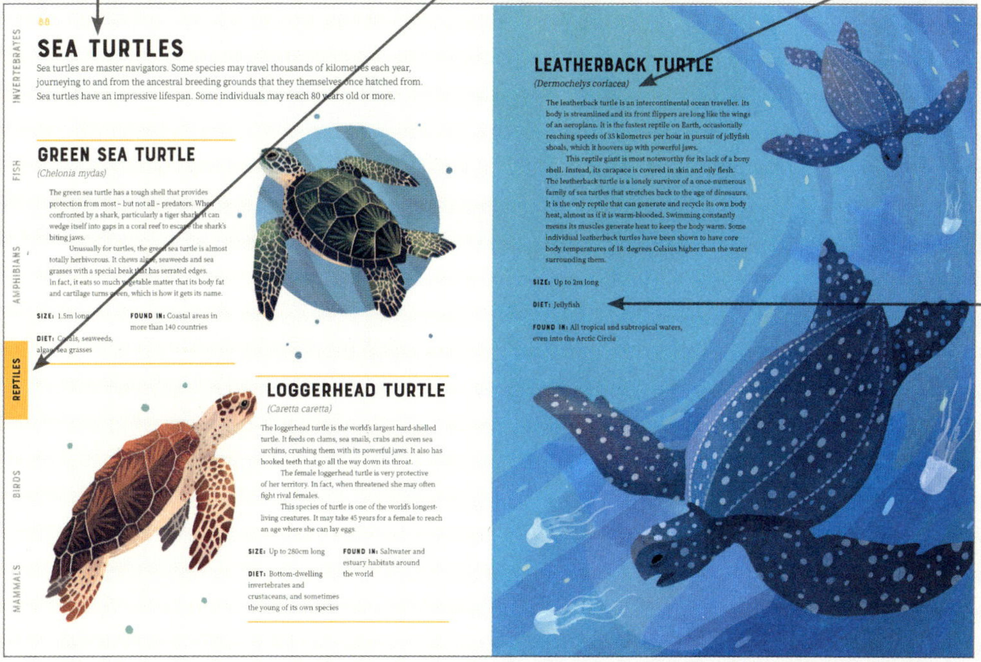

FACT FILE: Learn amazing facts about each creature here.

You can read this book in any order and in any number of sittings. Read it in your armchair, or take it out into the wild. Read about the creatures you know you love, and then discover many more you had no idea existed!

INVERTEBRATES

WHAT IS AN INVERTEBRATE?

If we could gather into a room one representative from each and every species of animal alive right now, nearly all of the animals you would see around you in that room would be invertebrates. You would see hundreds of thousands of different kinds of beetles, thousands upon thousands of flies, and fifty thousand or so different spider species.

Invertebrates | Everything else

All these animals have one thing in common: they lack a backbone or spine. In the corner of that imaginary room would be the animals *with* bones. Here, in the shadow of the bugs, bees and beetles, some familiar animals would sit: the fish, amphibians, reptiles, birds and mammals. On planet Earth, bony animals are in the minority. 95 per cent of all life on Earth has no backbone. These almost completely boneless animals are collectively known as invertebrates.

Invertebrates come in a dizzying array of shapes and sizes. Many species have hard exoskeletons and move around on jointed legs. These include the spiders, the scorpions, the insects and the crustaceans. Then there are the rather slimy invertebrates that often have shells, such as snails and clams – animals collectively called the molluscs. There are vast families of worms and wormlike invertebrates out there, many of which are awaiting discovery. The same goes for the jellyfish and other invertebrates such as the mites and the tiny rotifers and tardigrades.

Put simply, invertebrates are everywhere. In fact, we could not live without them for long. Invertebrates clean up dead animals; they deal with pests; they provide food for birds, reptiles, amphibians, mammals and fish; they recycle soil nutrients, keeping plants healthy; they pollinate the crops we eat and the vibrant and colourful plants we don't. Nature would be a boring place without invertebrates, as you will discover on the pages that follow.

INVERTEBRATE CHARACTERISTICS

SHEDDING SKIN
Many invertebrates, particularly arachnids, insects and crustaceans, shed their outer skin layer as they grow. This is a risky time for invertebrates because their new shell takes time to toughen up, making them vulnerable to predators. Many hide during this period.

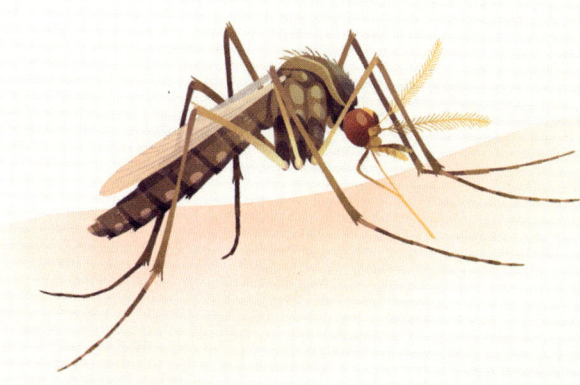

MOUTHPARTS
In vertebrates, most jaws simply go up and down to open and close. Invertebrates are armed with a range of specialist mouthpart adaptations that help these creatures undertake a variety of tasks. These include beaks used for pulling apart prey (octopuses), mouthparts that are used like syringes to suck blood (mosquitoes), jaws that slice (beetles), venomous fangs used to disarm prey (spiders) and tongues capable of scratching algae off surfaces (snails).

BABY STAGES
Many invertebrates, including most insects and crustaceans, hatch from eggs into larvae that differ from the adult form. Often these larvae live in different habitats to their parents and can look very different. In some cases, the larva becomes a pupa, within which it will completely transform into a new adult form. This is especially true of many flying insects, including moths and butterflies, whose larvae we call caterpillars.

REPRODUCTION
Some invertebrates can reproduce without the need for a partner. These include many species of snails and slugs which are able to fertilise their own eggs. In some species – most notably ants, wasps and bees – a single queen is responsible for an entire colony. She fills her colony with an army of sterile female workers, only producing male offspring later in the year.

WORMS AND WORMLIKE CREATURES

The term 'worm' refers to a mind-boggling array of creatures with long simple bodies, many of which tunnel through the ground feeding upon nutrients gathered from mouthfuls of sand and mud. However, not all worms and wormlike creatures occupy their habitats in this way. Some, including the Bobbit worm, are predators. Others, such as the medicinal leech, are parasites.

GIANT GIPPSLAND EARTHWORM
(Megascolides australis)

The giant Gippsland earthworm regularly reaches almost three metres in length, making it one of the longest earthworms in the world. It is also one of the longest-lived of all worms, with some individuals reaching an age of five years or more.

Few people ever get to see the giant Gippsland earthworm in real life because it hides in the soil and mud. To find this earthworm, scientists have to listen out for the characteristic slurping noise that wet mud makes as the worm squeezes its long body through a network of tunnels.

SIZE: Up to 3m long

DIET: Roots and tiny particles of food in soil

FOUND IN: Clay soils along stream banks in Victoria, Australia

MEDICINAL LEECH
(Hirudo medicinalis)

Compared to other leeches, the medicinal leech has impressive jaws. It has 100 teeth arranged onto three blades which can be stabbed into larger animals so that the leech can drink its meal – of blood. Special chemicals in its saliva make the bite painless so that bigger animals rarely know that the leech is there.

This leech species is so named because it was once regularly used in medicine. Doctors would attach leeches to patients to help them recover from illnesses. Today, the practice continues in some parts of the world.

SIZE: Up to 20cm long

DIET: Blood, including from mammals such as humans

FOUND IN: Isolated freshwaters across Europe and Asia

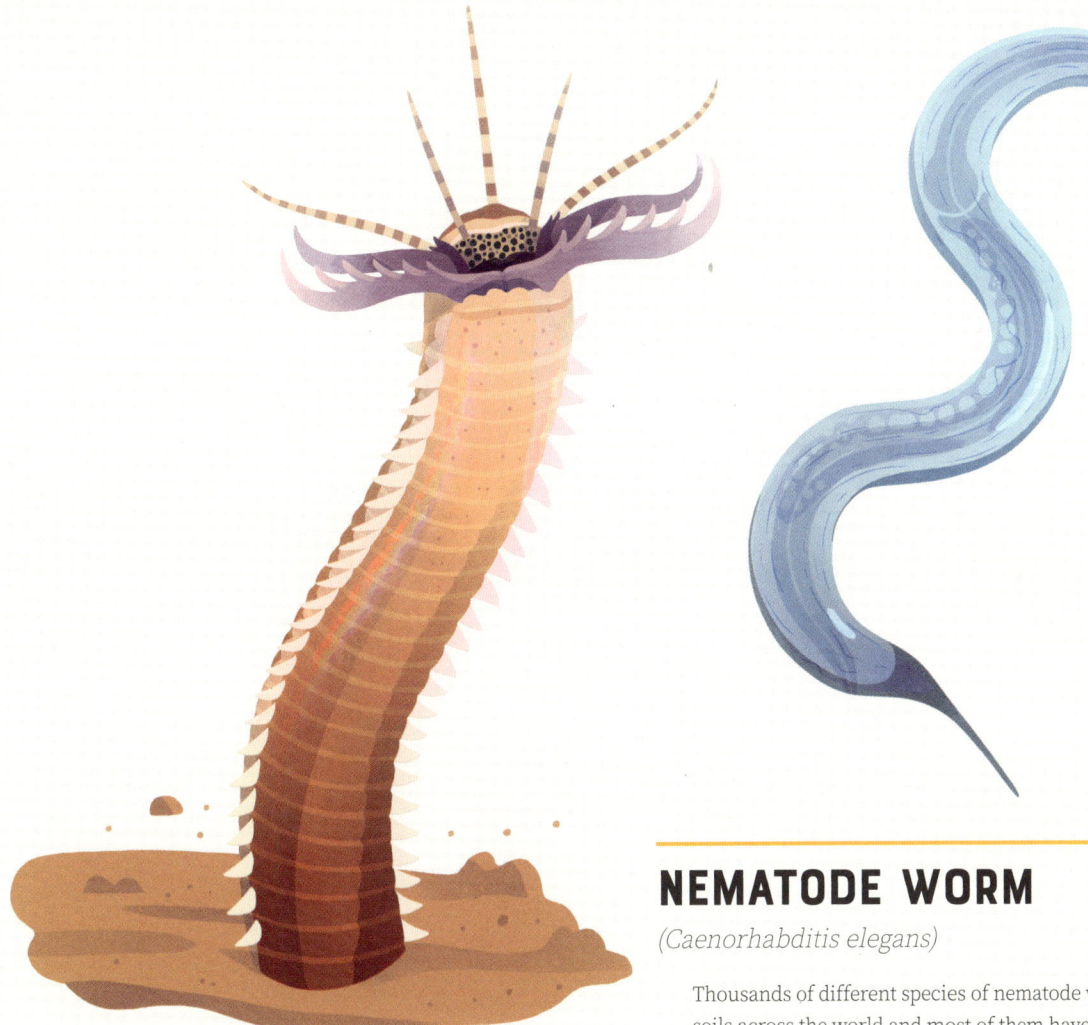

INVERTEBRATES

BOBBIT WORM
(Eunice aphroditois)

The Bobbit worm arms its trapdoor-like jaws and waits patiently in its burrow. If a fish should swim nearby and accidently brush one of its five antennae, the Bobbit worm switches on its kill response. In the blink of an eye, it lunges through the water and snaps its jaws shut.

These worms sometimes get into aquariums as accidental stowaways. In 2009, staff from one UK aquarium had to empty a tank in which a Bobbit worm (nicknamed Barry) was growing, because of its insatiable appetite for aquarium fish.

SIZE: Up to 3m long

FOUND IN: Warmer waters featuring gravels, muds and corals

DIET: Fish and other marine creatures

NEMATODE WORM
(Caenorhabditis elegans)

Thousands of different species of nematode worms live in soils across the world and most of them have not yet been named. Scientists, however, know this species of nematode worm very well.

Because it is easy to study with a microscope and can be bred in great numbers, *Caenorhabditis elegans* has helped scientists learn more about how animals work. By studying this worm, scientists are discovering why animals sleep, how (and why) animals age and how DNA (the building blocks of life) builds animal bodies.

SIZE: 1mm long

FOUND IN: Nutrient-rich soils all over the world

DIET: Bacteria

FISH AMPHIBIANS REPTILES BIRDS MAMMALS

OTHER WORMLIKE CREATURES

Worms can move around in a variety of ways. Some, such as the acorn worms, use their armoured head like a battering ram to force themselves through sand and mud. Others, like the flatworms, move using hundreds of tiny hairs that beat rhythmically, allowing them to skate on slime. And there are even worms that walk – the majestic and very ancient velvet worms.

PINK VELVET WORM
(Opisthopatus roseus)

The pink velvet worm is right on the edge of extinction. Its remaining habitat is a single forest in South Africa and it is rarely seen because it hides in logs and underneath rotting leaves.

Like all velvet worms, the pink velvet worm walks upon numerous rows of pairs of stilts, which are flexed by water pressure – almost as if it's walking on water balloons. The pink velvet worm hunts for invertebrates, which it catches by spraying a sticky glue-like substance that stops prey from running away.

SIZE: Up to 4mm long

FOUND IN: Weza Forest in South Africa

DIET: Invertebrates including termites and woodlice

ARROW WORM
(Parasagitta setosa)

Each and every morning, the arrow worm swims down into the deep sea where it hunts for microscopic prey that it pins down with eight or nine curved hooks near its mouth. It swims with the aid of fins, which make it look a little bit like a tiny fish. In a single day of hunting, *Parasagitta setosa* can eat almost its entire weight in tiny sea creatures. When the sun sets, it quietly floats back nearer the surface to avoid the attentions of predatory jellyfish.

SIZE: 14mm long

FOUND IN: Oceans throughout the northern hemisphere

DIET: Planktonic creatures including copepods and tiny sea squirts

PERSIAN CARPET FLATWORM
(Pseudobiceros bedfordi)

This attractive flatworm swims through the ocean by undulating (waving) the sides of its body like a flowing ribbon. Not all flatworms can move in this way. Many flatworms glide over rocks or mud using rows of tiny beating hairs greased up by a layer of slime.

As with many flatworms, every individual Persian carpet flatworm is both male and female. Individuals meet regularly near the sea floor where their energetic tussles lead to egg-laying in the days and weeks that follow.

SIZE: 8–10cm long

DIET: Sea squirts and small crustaceans

FOUND IN: Coral reefs throughout Southeast Asia and Australasia

ACORN WORM
(Ptychodera flava)

This species was one of the first ever so-called acorn worms discovered by scientists. It swallows mouthfuls of sand and pulls out the particles of nutrients from the sand as the sand passes through its gut. The undigested material comes out of its bottom as a big dropping, called a cast.

Like all acorn worms this one has a big rubbery protrusion coming out of its head. In *Ptychodera flava* this acts like a muscular shield, allowing it to pass easily through sand and mud.

SIZE: Up to 8cm long

DIET: Particles of nutrients in sand

FOUND IN: Tropical waters including near Australia, the Galapagos Islands, Hawaii and Mauritius

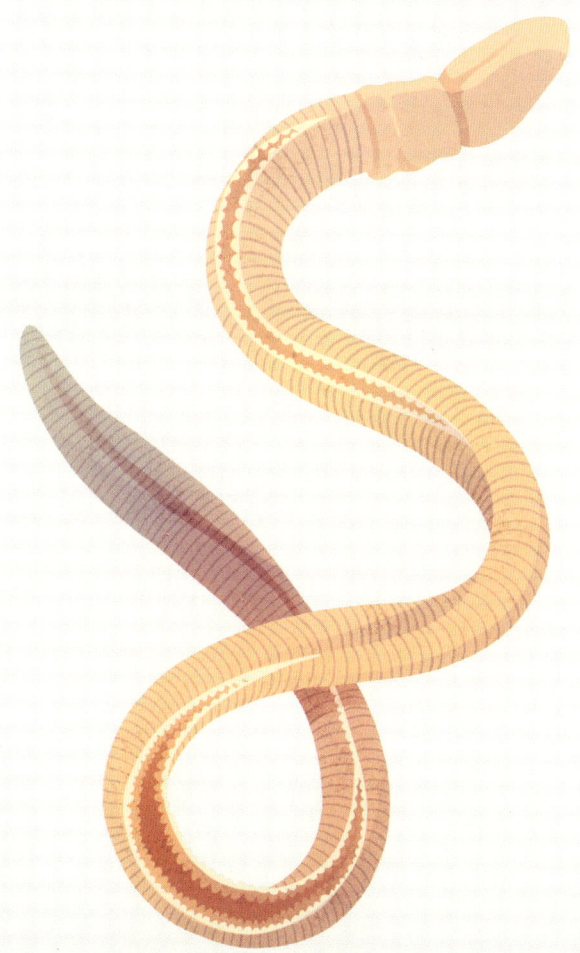

SEA JELLIES

Some of the most simple and effective life forms on Earth are the jelly-like invertebrates (loosely known as jellyfish) that make up the groups known as cnidarians and ctenophores. Many of these animals feed by trapping passing animals in their feeding tentacles.

SEA WASP
(Chironex fleckeri)

Many scientists consider the sea wasp to be the most lethal jellyfish species in the world. Between 1884 and 1996, at least 63 people died from its venomous toxins, which are among the most powerful known in nature. It delivers its deadly venom via millions of microscopic darts that cover its long, trailing tentacles.

The sea wasp belongs to a group of jellyfish known as box jellyfish. These are cube-shaped with big eyes and, though they lack a brain, they can manoeuvre around objects if needed.

SIZE: Up to 3m long

DIET: Prawns and small fish

FOUND IN: Coastal water throughout Australasia and Southeast Asia

SNAKELOCKS ANEMONE
(Anemonia viridis)

The snakelocks anemone gets its name from its long snakelike tentacles that resemble those of the snake-haired Medusa, a monster from Greek mythology. Its tentacles are slightly green because they are home to tiny algae which collect energy from the sun, helping power the anemone.

Unlike most jellyfish and anemones, the snakelocks anemone does not have babies that swim around like miniature jellyfish. Instead, its eggs are pumped into the water and settle on surrounding rocks, where they grow into their adult form.

SIZE: 8cm wide

DIET: Sea-snails and slugs, small fish

FOUND IN: Eastern coastlines of the Atlantic Ocean

SEA GOOSEBERRY
(Pleurobrachia bachei)

The sea gooseberry belongs to a primitive group of sea creatures called ctenophores that move by beating tiny rows of hairs called cilia. Rather than catching prey using thousands of venomous barbs like a jellyfish does, the sea gooseberry uses a pair of sticky tentacles, which it can flare out alongside the body like a pair of spider webs. When tiny prey becomes trapped, the tentacles are pulled back into the body and digestion begins.

SIZE: Body 2cm long with trailing tentacles up to 15cm

FOUND IN: West coast of North and Central America

DIET: Tiny crustaceans and other planktonic creatures

PORTUGUESE MAN O' WAR
(Physalia physalis)

Unlike nearly all other animals on Earth, each Portuguese man o' war is not one animal but many. Like a floating skyscraper, its body is made up of an army of single-celled organisms, called zooids, that are like the bricks that make up the body and the tentacles. There are zooids for catching prey, for digesting prey and for manufacturing eggs.

The Portuguese man o' war cannot swim. Instead it rides ocean currents using an air-filled life-jacket with a sail that can catch the wind.

SIZE: Floating chamber 9–30cm long, with tentacles up to 30m long

FOUND IN: Throughout the Atlantic Ocean, Indian Ocean and Pacific Ocean

DIET: Small marine animals including small fish and plankton

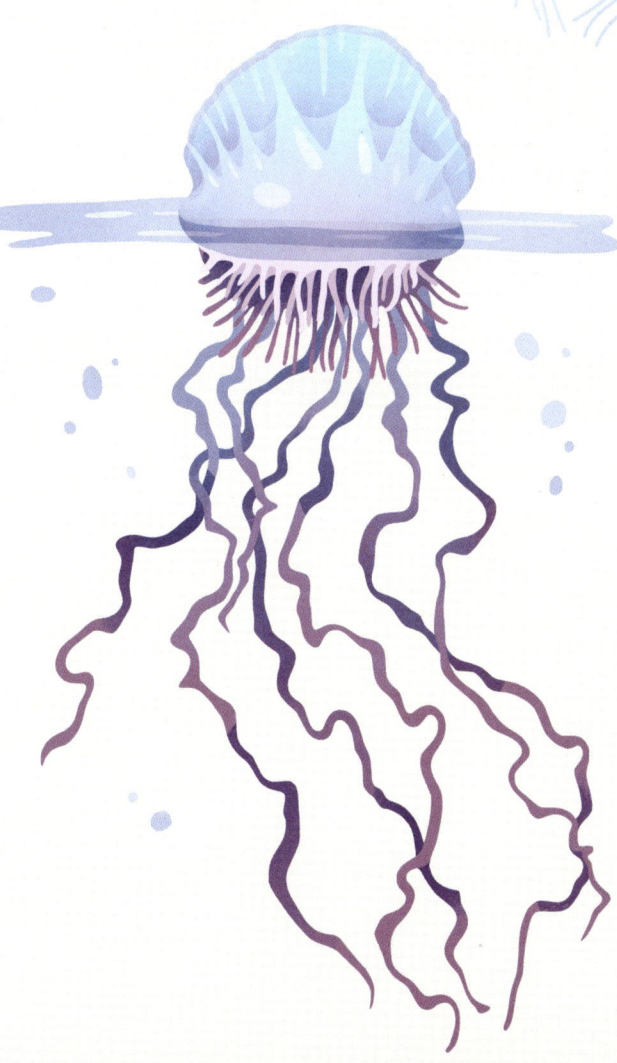

SPONGES AND CORALS

Some of the most ancient animals of all, which have lived on Earth since the earliest prehistoric times, are the sponges and reef-building invertebrates such as corals. In the case of corals, their skeletons provide the architecture for an undersea cityscape used each day by sea snakes, sharks, rays and countless other species of fish.

BATH SPONGE
(Spongia officinalis)

This simple animal sieves tiny particles of food from the water, which it funnels through its basket-like body. It grows incredibly slowly, taking as long as 40 years to grow to the size of an apple.

The bath sponge is found throughout the Mediterranean Sea. It was once harvested to near extinction because humans considered it a handy tool with which to scrub and wash during bath time. Today most sponges for human use are artificially produced, though some species are still farmed for shops to sell.

SIZE: Up to 35cm in diameter

FOUND IN: The Mediterranean Sea and Caribbean Sea

DIET: Bacteria and other food particles in the water

GROOVED BRAIN CORAL
(Diploria labyrinthiformis)

Each night, thousands of tiny tentacles emerge from colonies of the grooved brain coral to feed upon passing plankton. Many of these tentacles come out from special channels found across the surface of the coral structure. These trenches make this coral resemble a giant human brain, hence the name.

Like sponges, many corals produce a skeleton-like structure as they grow, which offers them support and protection. These skeletons are produced very slowly. It might take 500 years for a brain coral colony to reach its largest size.

SIZE: Up to 2m

FOUND IN: Western Atlantic Ocean and Caribbean Sea

DIET: Tiny zooplankton and bacteria

MICROSCOPIC ANIMALS

Some of the weirdest invertebrates are those that are so small we cannot see them without the aid of a microscope. They include some of the most under-studied invertebrates out there, including the rotifers and the tardigrades. Countless other undiscovered species like those on this page are out there for a new generation of scientists to discover one day.

TARDIGRADE
(Milnesium tardigradum)

The microscopic *Milnesium tardigradum* represents a very strange group of animals called tardigrades. These tiny creatures can live almost anywhere, from mud volcanoes to tropical rainforests and even in the icy wastelands of Antarctica. They thrive in their thousands upon mosses and lichens.

In 2007, scientists took this species to space to see if it could survive the sub-zero temperatures and the immense vacuum of space. Incredibly, some individuals survived unharmed, making this one of the toughest animals on the planet.

SIZE: Up to 0.7mm long

FOUND IN: Numerous land habitats all over Earth

DIET: Smaller tardigrade species plus algae and tiny worms

ROTIFER
(Rotaria rotatoria)

Rotifers are a large group of common microscopic animals, each of which possesses a wheel-like mouth covered in rows of tiny beating hairs. These hairs create a miniature vortex in the water which draws food particles ever closer to the mouth. As with tardigrades, many rotifers can shrivel up their bodies and become a seedlike husk able to survive years without water.

Species such as *Rotaria rotatoria* are almost unique on Earth because all individuals are females. Scientists think that no males of these rotifers have existed for millions of years.

SIZE: 0.2–1.5mm long

FOUND IN: Ponds, lakes and streams

DIET: Bacteria, algae and other freshwater detritus

STARFISH AND RELATED CREATURES

The invertebrates on this page belong to a family group known as the echinoderms. Each echinoderm species has five-way symmetry, which gives many of them an obvious star-shaped appearance. Echinoderms were one of the first animal success stories in the world's oceans 500 million years ago. With 7,000 species alive today, they are still doing very well.

SUNFLOWER SEA STAR
(Pycnopodia helianthoides)

The sunflower sea star is one of the world's toughest sea stars. It has a meshlike skeleton that acts like a protective cage covering its internal organs. Powerful suckers can grip onto coral during the heaviest storms so that it can stay safe.

The sunflower sea star has between 16 and 24 legs, inside which are thousands of tube feet that allow it to move at the impressive pace (for a starfish) of one metre per minute.

SIZE: Up to 1m in diameter

DIET: Mostly sea urchins

FOUND IN: Northeast Pacific Ocean, from California to Alaska

FLOWER URCHIN
(Toxopneustes pileolus)

What look like flowers on this urchin are actually numerous colourful grabbers capable of stabbing at predators with venom-tipped fangs. The flower urchin has hundreds of these grabbers all over its body, making it an animal most predatory fish leave well alone.

Unusually, the flower urchin is capable of dressing itself. It uses various sucker-like appendages to pull rocks and stones against its body to create a heavy armour that may help stop it from drifting away during storms.

SIZE: Up to 16cm in diameter

DIET: Algae and ocean detritus

FOUND IN: Coral reefs across many parts of the Indian Ocean and Pacific Ocean

SNAKE SEA CUCUMBER
(Synapta maculata)

Unlike other sea cucumbers, the snake sea cucumber doesn't spray its guts out at passing predators when scared. Instead, it drops unnecessary parts of its body onto the sea floor to keep predatory fish occupied, and then makes a quiet escape.

Like their starfish cousins, sea cucumbers have a glasslike skeleton just underneath the surface of their skin. A sea cucumber's mouth is surrounded by tentacles covered in feather-like blades that feel around the sea floor searching for its favourite food, rotting fish and sea grasses.

SIZE: Up to 3m long – this is the largest sea cucumber

FOUND IN: Shallow waters of the tropical Indo-Pacific Ocean

DIET: Sea grasses and decomposing matter

COMMON BRITTLESTAR
(Ophiothrix fragilis)

The legs of the common brittlestar are incredibly fragile and fall off very easily. When near the crashing waves at the coast, it must keep hidden under rocks and sometimes even seashells to stay safe. It pulls itself across the ocean floor using two of its legs at a time.

Further out to sea, common brittlestars can group together in enormous numbers. In some parts of the sea floor there can be more than 2,000 of them per square metre.

SIZE: Up to 22cm across

FOUND IN: Eastern coast of the Atlantic Ocean and the North Sea

DIET: Scavenger of dead animals though also capable of filter-feeding

SHELLFISH

Many invertebrates make use of a pair of hard shells (valves) which protect them from predators and from drying out when not in water. This adaptation is commonly seen in filter-feeding molluscs called bivalves. One other group (called brachiopods) have hit upon the same armoured lifestyle. Of the animals shown here, the lamp shell is an example of a brachiopod.

PACIFIC RAZOR CLAM
(Siliqua patula)

The Pacific razor clam is a masterful digger. By shooting salty water downwards into the soft shoreline it creates a depression in the sand which it can force its big round muscular foot into. By doing this again and again it burrows downwards, digging out a tunnel in which it rests. When the tide comes in, the razor clam unveils special feeding tubes called siphons. These tubes hoover up plankton in the water.

SIZE: Over 16cm in length

DIET: Plankton

FOUND IN: Sandy unpolluted beaches of the North American west coast

GIANT CLAM
(Tridacna gigas)

By pulling water into its vast shells, the giant clam is able to filter out particles of food which it uses to grow. But the giant clam has a secret. Within its cells are tiny algal hitch-hikers which get their energy from sunlight. The giant clam harvests some of their energy as an additional food source, helping power its supersized growth.

Historical specimens of giant clam were very large. Some weighed almost a third of a tonne.

SIZE: Up to 120cm across

DIET: Microscopic plankton and other organic matter

FOUND IN: South Pacific and Indian Ocean

BAY SCALLOP
(Argopecten irradians)

Scallops are very wary bivalves. Should any of their numerous eyes spot an approaching predator, they clap their shells together like maracas to create a simple means of jet propulsion that allows them to escape.

The bay scallop used to be very common 50 years ago but their numbers have dropped dramatically. Scientists think this could be because there are fewer sharks due to overfishing, which means sharks' prey animals like rays are flourishing. Rays are common predators of the bay scallop.

SIZE: 55–90mm

DIET: Plankton, which it filter-feeds through its gills

FOUND IN: Across the Atlantic coastline of North America

LAMP SHELL
(Lingula anatina)

Lamp shell burrows can be spotted by looking for three holes in the wet sand. The holes at either end are where food-rich water is pulled in to the lamp shell's mouth, and the middle hole is where the water is pushed out.

Lamp shells and their brachiopod relatives once rivalled the bivalves for world domination, but their rise to power was cut short by an enormous extinction event that hit the Earth 250 million years ago. Since this time, the bivalves have outcompeted brachiopods in most parts of the planet.

SIZE: Shell between 5 and 6cm

DIET: Plankton

FOUND IN: India's coastlines and estuaries

GASTROPODS

Gastropods (otherwise known as slugs and snails) are the most numerous group of a category of animals called molluscs. In all, scientists estimate there are between 65,000 and 80,000 species of gastropod alive today. They range from colourful sea slugs such as the Spanish dancer to terrestrial success stories like the slugs and snails so common in our gardens.

LEOPARD SLUG
(Limax maximus)

While most humans are in bed, the leopard slug undergoes an amazing ritual in gardens. Each individual slug must find another slug, whom it will spend hours circling and chasing. If the two slugs find one another attractive enough, they create a bungee cord of slime and they dangle off a tree branch. Here, swinging metres off the ground, they fertilise one another's eggs.

The leopard slug can be a very effective cleaner-up of dead creatures in urban areas. While we sleep, many dead rats, pigeons and frogs are feasted upon by these hungry molluscan vultures.

SIZE: 10–20cm long

FOUND IN: Gardens throughout Europe

DIET: Plants, fungi and other slugs, as well as decaying vegetation and animals

SPANISH DANCER
(Hexabranchus sanguineus)

By day, the Spanish dancer waits silently in cracks and crevices in the coral reef. By night, it glides across the ocean floor seeking out sponges on which to feed. If it is disturbed by predators during these hunting forays, the Spanish dancer lifts up the edges of its body and swims rhythmically through the water as if to hypnotise its challenger. This undulating swimming style looks a bit like a flamenco dancer's skirt, which is how this sea slug gets its common name.

SIZE: Up to 60cm long

FOUND IN: Coral reefs throughout many parts of the Indian and Pacific Oceans

DIET: Sponges

GIANT AFRICAN SNAIL
(Achatina achatina)

The giant African snail dwarfs nearly all other snail species. Its huge heavy shell and slimy muscular body can weigh almost a kilogram in some individuals, which is five times heavier than a hamster.

Giant African snails are a traditional food source for some people in West Africa, where the snails are often gathered from the wild or farmed in 'snaileries'.

Being both male and female, the giant African snail can breed very quickly. Each year, one individual can produce as many as 1,200 eggs. Its ability to have lots of babies makes it a pest species if it is introduced to countries where it doesn't belong. This can happen when pet giant African snails escape and move into the wild. Because they have no natural enemies, it is very difficult to control their spread.

SIZE: Up to 30cm long

DIET: Leaves, grasses, fallen fruits and vegetables

FOUND IN: Forests and grasslands of West Africa

CEPHALOPODS

The cephalopods are the cleverest molluscs known to scientists. This group includes the octopuses, cuttlefish, squids and the shelled octopus-like creatures such as the nautilus. In all, more than 800 cephalopod species have been described.

GREATER BLUE-RINGED OCTOPUS
(Hapalochlaena lunulata)

When first approached by predators, the greater blue-ringed octopus can change the colour and texture of its skin to camouflage itself against its surroundings. If this doesn't work, it flashes vibrant patterns of yellows and blues at its predator to scare it away. If this fails too, the octopus must rely on a highly venomous bite. Bites to unwary humans can result in a total body paralysis that leads to suffocation. Thankfully, human deaths from this cause are rare.

SIZE: 10cm across including tentacles

DIET: Crustaceans, bivalves and small fish

FOUND IN: Shallow subtropical and tropical waters across the Indo-Pacific Ocean

ATLANTIC GIANT SQUID
(Architeuthis dux)

For many years, this deep-sea squid had almost mythical status as we knew so little about it. Now, through the use of submarines with special cameras, scientists are beginning to learn more about its secretive way of life.

While hunting, the giant squid can hang motionless in the water, its enormous eyes primed to spot the movement of any prey swimming nearby. Using a pair of extra-long tentacles armed with serrated sucker rings, the giant squid strikes. Prey is pulled apart with help from a razor-sharp beak. This squid is also highly mobile. Like all cephalopods, it swims using jet propulsion.

SIZE: Up to 13m including its two feeding tentacles

FOUND IN: Deep seas throughout the world

DIET: Fish and small squids, including those of its own species

CHAMBERED NAUTILUS
(Nautilus pompilius)

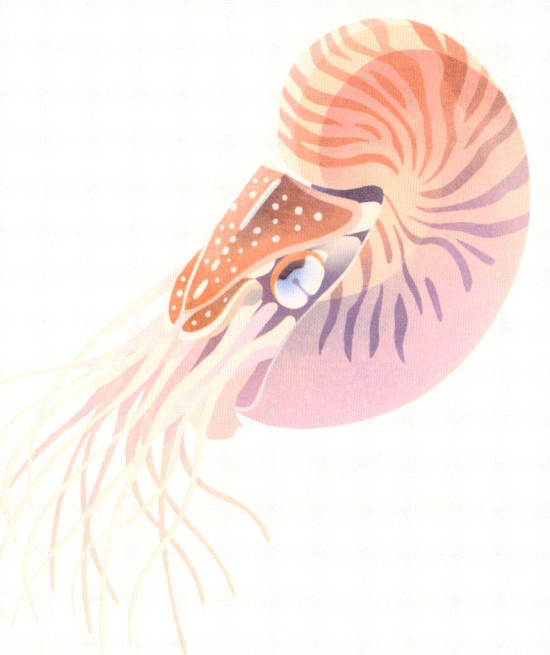

Unlike most cephalopods, the chambered nautilus has a protective shell that can be filled or emptied with gases to help move up or down in the water. Guided by a strong sense of smell, it moves throughout rocks and coral reefs looking for decaying food. Its eyes are very primitive.

Unlike other cephalopods, nautilus babies do not have a free-swimming larval stage that drifts around as plankton. Instead, baby nautilus are born with shells and immediately look like tiny versions of adults.

SIZE: Up to 20cm long

DIET: Crabs, shellfish and dead animals

FOUND IN: Reefs and rocky outcrops in oceans from Japan to Australia

COMMON CUTTLEFISH
(Sepia officinalis)

The common cuttlefish is one of most expressive animals on Earth. Through special organs in its skin, combined with sheets of mirror-like structures on its surface, it can change both the colour and texture of its body to communicate with those around it. Some cuttlefish patterns express anger or rage whilst other patterns are used to show off to potential love interests.

The common cuttlefish often changes colour to blend into its surroundings. Many cuttlefish use this trick to hide amongst the sand where they wait to ambush unwary prey.

SIZE: Up to 45cm long not including the tentacles

DIET: Crustaceans, small fish and molluscs (including smaller individuals of its own species)

FOUND IN: Sandy and muddy sea floors across the Mediterranean Sea, North Sea and Baltic Sea

CRUSTACEANS

Crustaceans are a large group of arthropods – invertebrates with a segmented body and pairs of jointed limbs. Unlike insects, crustaceans have double-branched legs and a variety of strange larval forms. Some have evolved hard shells for protection. They manage to live in some of the toughest places on Earth. There are sea species, freshwater species and even some land crustaceans, such as woodlice.

HALLOWEEN CRAB
(Gecarcinus quadratus)

As long as its deep burrow remains damp, the Halloween crab can survive quite happily away from water, though it must return to the sea each year to lay eggs. At night-time, when the ground is wet, it emerges from its large burrow to look for scraps to feed on.

Land crabs like this can be very helpful to the environment because their tunnelling mixes up rainforest soils and helps plants grow. For this reason, some scientists refer to them as 'ecosystem engineers'.

SIZE: 5cm long (not including legs)

DIET: Leaf litter and seedlings

FOUND IN: Common in the coastal rainforests of Central America

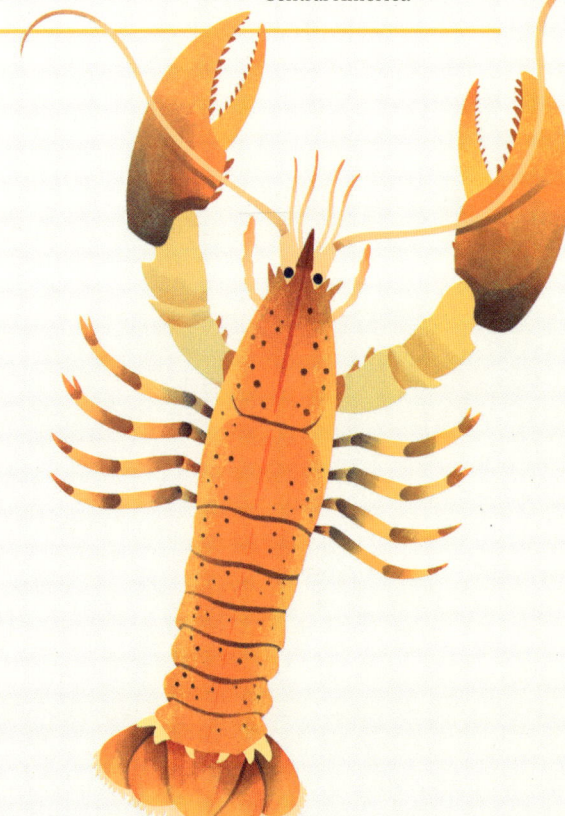

AMERICAN LOBSTER
(Homarus americanus)

Unlike in the case of most invertebrates, the left and right claws of the American lobster differ from one another. One claw is larger and has special rounded edges to crush prey. The other claw is smaller and sharper and is used for holding or pulling apart prey.

Weighing three times as much as a bowling ball, the American lobster is the heaviest crustacean alive today. It may also be one of the most long-lived of invertebrates. Some individuals may live 100 years or more.

SIZE: 64cm long not including claws

DIET: Mostly mussels, starfish and large worms

FOUND IN: Cold shallow waters across much of the Atlantic coast of North America

PEACOCK MANTIS SHRIMP
(Odontodactylus scyllarus)

Pound for pound, nothing in nature packs a punch quite like that of the peacock mantis shrimp. Using special fistlike appendages it delivers its strategic blows at more than 80 kilometres per hour with an explosive force equivalent to that of a bullet firing from a gun. These powerful punches are often more than enough to shatter the protective armour of its shellfish prey.

When not hunting, the peacock mantis shrimp spends much of its time hiding in a U-shaped hole dug into the sand beneath the coral reef.

SIZE: Up to 18cm long

DIET: Molluscs and crustaceans

FOUND IN: Coral reefs across many tropical parts of the Indo-Pacific Ocean

DESERT WOODLOUSE
(Hemilepistus reaumuri)

Like all woodlice, the desert woodlouse walks across land on seven pairs of jointed legs. To stay alive in such a dry place, this crustacean must obtain all of its water from the plants on which it feeds and by recycling moisture from the air. To avoid wasting water, it sleeps in family groups squashed together in tight burrows. These families are highly territorial. Each family group marks its property using smelly droppings that ward off neighbouring families.

SIZE: Up to 22mm long

DIET: Leaves

FOUND IN: Dry deserts of North Africa and the Middle East

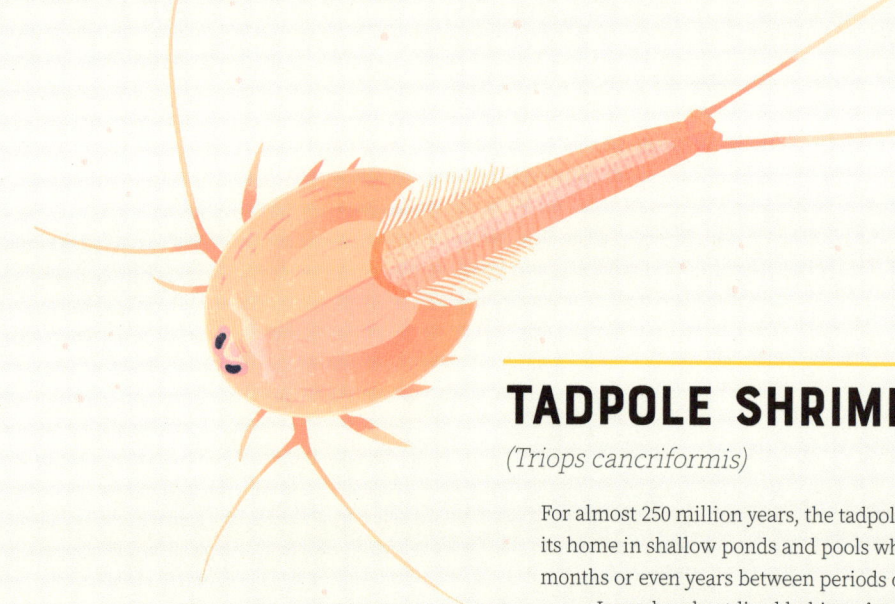

TADPOLE SHRIMP
(*Triops cancriformis*)

For almost 250 million years, the tadpole shrimp has made its home in shallow ponds and pools which dry up for months or even years between periods of heavy rain.

In such a short-lived habitat, the tadpole shrimp has to grow incredibly quickly. After hatching, it manages its journey to adulthood in just three weeks, before laying its own eggs which come to rest on the bottom of the pond as the habitat dries up again. These eggs can survive in the dry mud for more than 25 years. When water returns, its curious life cycle begins anew.

SIZE: 11cm long (shell)

DIET: Small invertebrates, aquatic plants and mud

FOUND IN: Isolated temporary ponds and pools throughout Europe

GODZILLA REMIPEDE
(*Godzillognomus schrami*)

The Godzilla remipede was first discovered by intrepid scuba divers exploring remote underwater caves. In these pitch-black habitats, it hunts for prey by feeling for the movements of small invertebrates in the water. Like other remipedes, the Godzilla remipede latches onto prey using fangs that can inject poison. This makes it one of the world's only venomous crustaceans.

For many years, remipedes were only known from fossils but today scientists know of 28 different living species. Of all the crustaceans, this ancient group is perhaps most closely related to the insects.

SIZE: 7mm long

DIET: Small swimming invertebrates

FOUND IN: A single underwater cave on Eleuthera Island in the Bahamas

PACIFIC GIGANTIC SEED SHRIMP
(Gigantocypris agassizii)

Deep in the ocean's darkest depths, the Pacific gigantic seed shrimp motors around using its antennae like the oars on a boat. When threatened by predators it retreats into its shell-like body and drifts silently away from trouble.

When hunting, this shrimp watches out for the bioluminescent glow that its tiny prey give off as they move. Its eyes contain lots of special mirrors that help it detect these tiny bursts of light from many directions at once.

SIZE: Up to 3.2cm in diameter

FOUND IN: Deep parts of the Pacific Ocean

DIET: Tiny ocean invertebrates and baby fish

POLI'S STELLATE BARNACLE
(Chthamalus stellatus)

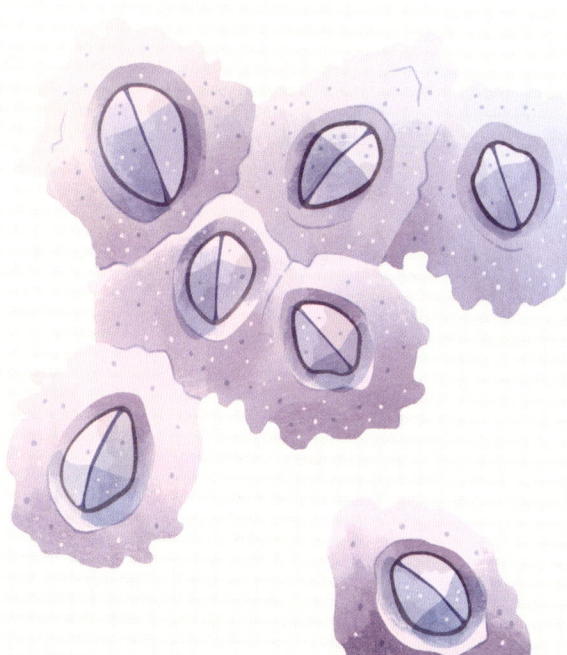

At first glance the Poli's stellate barnacle looks like a clam or an oyster, yet within its hard shell is a familiar crustacean body plan. It uses its long, feathery legs to sieve particles of food from out of the water.

Like many other sea-living crustaceans, the Poli's stellate barnacle produces thousands of tiny larvae that swim through the ocean amongst plankton. These larvae drift in ocean currents before gluing themselves to rocks to begin life as filter-feeding adults.

SIZE: Up to 14mm in diameter

FOUND IN: Rocky shores in the UK and southern Europe

DIET: Plankton and tiny food items in water

SPIDERS

Spiders are one of the most diverse arthropod groups on Earth. More than 51,000 species are known; each has eight legs and a single pair of venom-filled fangs. Spiders are important predators of insects. Each year, the world's spiders consume as much as 800 million tonnes of insect matter.

GOLIATH BIRDEATER
(Theraphosa blondi)

Weighing almost as much as a hamster, the Goliath birdeater is the heaviest spider in the world. Despite its name it mostly feeds on other invertebrates, although frogs, salamanders and even small snakes can be preyed upon too. It spends most of the day in deep underground burrows, coming out only at night to hunt.

Like other tarantulas, the Goliath birdeater can use its fangs to deliver a venomous bite if threatened or cornered by predators. Thankfully, this bite is rarely life-threatening to humans.

SIZE: Up to 11.9cm long (body) with a legspan up to 28cm

FOUND IN: Swamps and marshy areas throughout tropical South America

DIET: Large insects, spiders, worms and amphibians

DARWIN'S BARK SPIDER
(Caerostris darwini)

The Darwin's bark spider is a record-holder for producing the toughest material known from nature. Each individual thread of its silk is 10 times stronger than Kevlar, the material used in bulletproof armour. It uses its extra-strong silk to build vast webs across rivers so that it can catch invertebrates flying near the water surface.

As with many spiders, the female Darwin's bark spider eats the male spider after mating. Scientists think this might give her an extra energy boost to help produce extra-healthy eggs.

SIZE: 6–18mm long (body)

DIET: Mayflies and other flying insects

FOUND IN: Tropical wetlands in Madagascar's Andasibe-Mantadia National Park

PEACOCK JUMPING SPIDER

(Maratus volans)

The peacock jumping spider gets its name from the vibrant and reflective patterns on the male's abdomen. If a female peacock jumping spider shows an interest, the male spider lifts up his abdomen and flattens it out to show off his range of colours whilst furiously waving his legs in an elaborate dance. Some males can keep this energetic dance up for 50 minutes or more. This curious behaviour helps females work out which male might make the best father for her offspring.

SIZE: 5mm long (body)   **FOUND IN:** Across isolated parts of Australia

DIET: Crickets and spiders

CAROLINA WOLF SPIDER

(Hogna carolinensis)

Deep in its underground burrow, the Carolina wolf spider waits for the sun to set. When night-time falls it scans the entrance of its burrow with large, sensitive eyes. If there are no predators watching, it comes out to hunt. The Carolina wolf spider is one of the largest of the wolf spiders. Like other wolf spiders, it is a solitary, athletic hunter with powerful legs. Female wolf spiders are unusual for using their silk to weave special egg sacs which they carry around on their backs like a little backpack.

SIZE: 19–25mm long (body)   **FOUND IN:** Throughout the USA

DIET: Grasshoppers and crickets

SPIDER-LIKE CREATURES

The spiders are part of a mighty group of arthropods called the arachnids. There are other spider-like creatures in this group. Nearly all arachnids are hunters and most use eight legs for walking, although in some species, such as the tailless whip scorpions, two of the legs are used as feelers.

TANZANIAN TAILLESS WHIP SCORPION
(Damon variegatus)

The Tanzanian tailless whip scorpion stands motionless with its pair of spiky grabbers primed and ready to lunge at unwary prey.

Like other tailless whip scorpions, this species moves on only three of its four pairs of legs. It uses its first pair of legs as long feelers to help it move around in dark places.

So-called whip scorpions are neither spiders nor true scorpions but, rather, a close cousin of both of these arachnid groups. Unlike many arachnids, they have no venom and cannot produce silk.

SIZE: 20cm legspan

DIET: Crickets and other large insects

FOUND IN: Often found under stones and logs in Tanzania and Kenya

EMPEROR SCORPION
(Pandinus imperator)

Unlike with many scorpions, the tail stinger of the emperor scorpion does not deliver a potent venom. Its effect on humans is a little bit like that of a bee sting. Instead, the emperor scorpion relies on very powerful pincers to trap and pull apart its prey. Adults also use these claws like a battering ram to push through the armoured walls of termite mounds.

The emperor scorpion is a giant among arachnids. It is rarely spotted because it is covered in sensitive hairs that detect the movement of approaching humans, helping it to flee out of sight.

SIZE: 20cm long

DIET: Insects including termites, and sometimes larger prey such as lizards and mice

FOUND IN: Savannahs and rainforests throughout tropical Africa

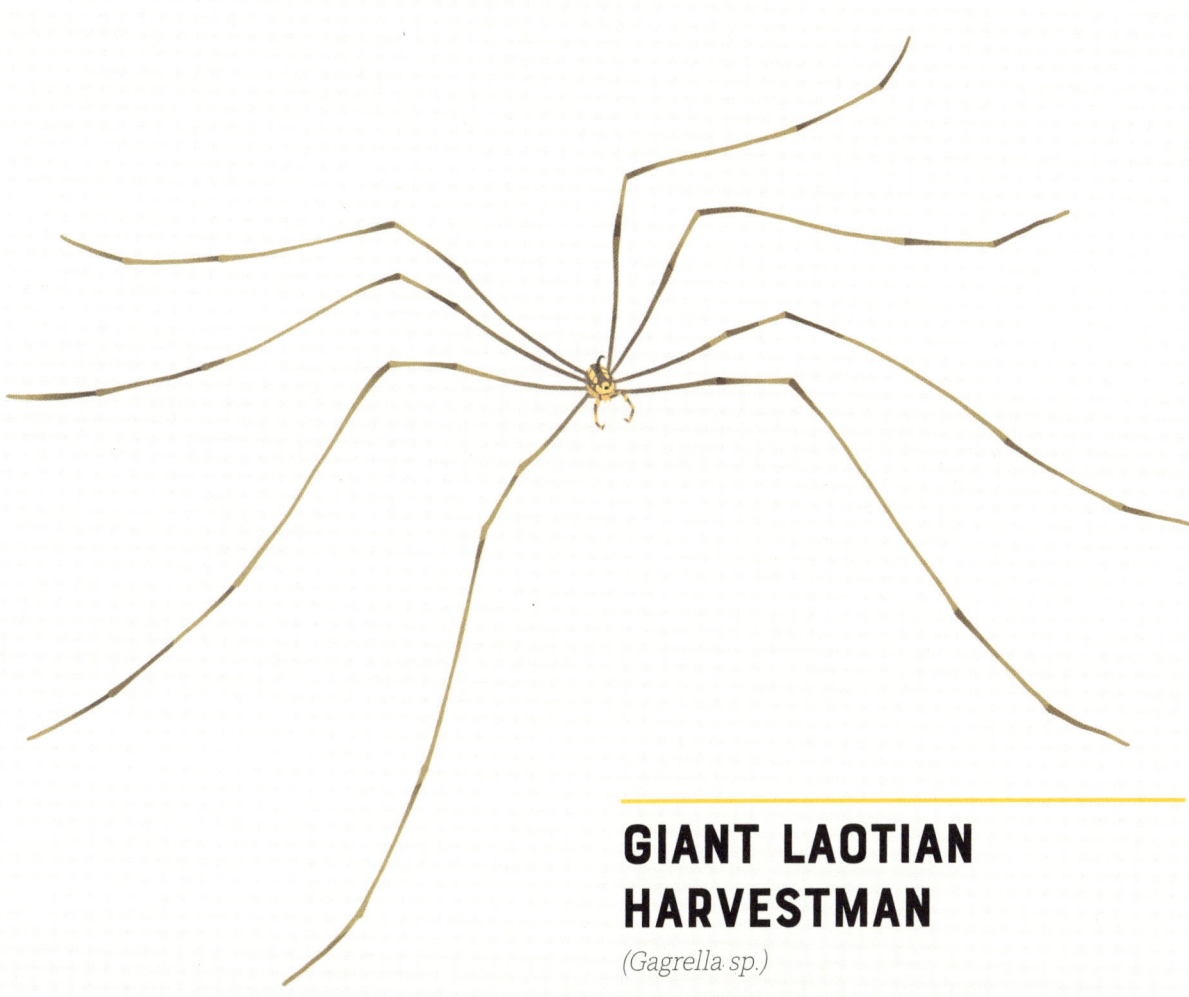

GIANT LAOTIAN HARVESTMAN
(Gagrella sp.)

No one knows quite why the giant Laotian harvestman became so big, but the mystery may have something to do with its long legs. Because harvestmen breathe through their legs, it may be that having extra-long legs means that this species can survive in habitats that are low in oxygen, such as caves.

The giant Laotian harvestman was only discovered in 2012. It is one of 6,650 harvestman species currently known to scientists but there are likely to be many more out there awaiting discovery.

SIZE: 33cm legspan

DIET: Likely to be omnivorous, feeding on decaying plants and animal droppings

FOUND IN: Remote caves of Laos

HYMENOPTERA

The largest group of arthropods is the insects, and hymenopterans, a group which includes the ants, bees and wasps, are a very important part of the insect family tree. In most species, females have a distinctive tail-spike (called the ovipositor) through which eggs can be delivered into hard-to-reach places, including into other animals. In many species, this spike can be used to deliver a painful sting.

WESTERN HONEY BEE
(Apis mellifera)

When the western honey bee finds nectar-filled flowers, it returns to the nest and tells others information about the flower's distance and its direction by performing a special tail-wagging dance in the hive. This makes it a master of animal communication.

The western honey bee was one of the first insects to be domesticated by humans, who make wooden hives for bee colonies, from which honey can be easily gathered. Today, the species has become one of the world's most important insects because of its skill for pollinating agricultural crops.

SIZE: Worker bees 10–15mm long, queens 18–20mm long

FOUND IN: Every continent on Earth except Antarctica

DIET: Nectar

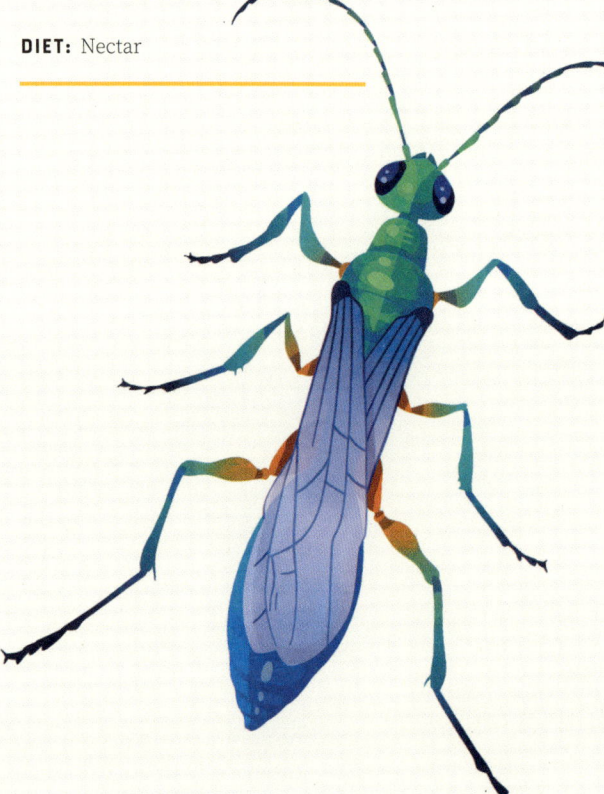

EMERALD COCKROACH WASP
(Ampulex compressa)

The female emerald cockroach wasp is on the lookout for a cockroach, for which she has a ghoulish plan. When she finds a suitably sized cockroach, she carefully delivers just the right amount of venom with her stinger to turn her prey into a walking zombie. She then leads the zombie cockroach back to her burrow where a single baby wasp will grow as a parasite inside its body. This parasitic behaviour may sound chilling, but the actions of wasps like these can help keep pest species in check.

SIZE: 22mm long

DIET: Cockroaches

FOUND IN: Tropical parts of South Asia, Africa and the Pacific islands

ARGENTINE ANT
(Linepithema humile)

At any given moment there may be as many as one million billion ants walking around on Earth. Their combined weight might even come close to the weight of all of humans alive today.

The secret to ants' success is their highly complex social networks. In many species, their nests are run by egg-laying queens, who are tended to by an army of 'daughters' who cannot themselves reproduce. The daughters are called worker ants. Worker ants are excellent at finding food, much of which they give to the larvae.

The Argentine ant is the most successful of all ants. Its large and complicated colonies often split into breakout colonies, which move off and set up new homes nearby. This spreading behaviour can lead to the setting-up of enormous 'super-colonies' many miles wide. The largest known super-colony of Argentine ants is along Europe's Mediterranean coast. Here, a single super-colony spans an incredible 6,000 kilometres of coastline.

SIZE: 2–3mm long

DIET: Omnivorous

FOUND IN: Originally from South America but has been accidentally released into many countries where it has become an invasive species

INVERTEBRATES

FISH

AMPHIBIANS

REPTILES

BIRDS

MAMMALS

BEETLES

If you were to gather every animal species on Earth and arrange them into one long line, one out of every four species would be a beetle. The key to the success of this enormous insect group is the hard exoskeleton, which offers armour-like protection and a pair of long wings which can be unfolded and deployed when needed.

ROYAL GOLIATH BEETLE
(Goliathus regius)

Weighing more than twice as much as a mouse, the Royal Goliath beetle is one of the world's largest insects. Its shell-like body offers it protection from a host of predators, including birds. Beneath its tough wing-cases are a beautiful pair of wings which allow it to move with agility from tree to tree.

Like many beetles, the larvae of this species live in soil where they can reach a length of more than 130 millimetres. When the dry season comes, these larvae pupate, ready to emerge as adults during the rainy season when food is especially plentiful.

SIZE: 50–110mm long (adult)

FOUND IN: Tropical rainforests across western Africa

DIET: Tree sap and fruits

BIG DIPPER FIREFLY
(Photinus pyralis)

Fireflies are actually beetles. In all, scientists have discovered more than 2,100 species, each able to generate varying amounts of light to show off to one another on dark nights. Fireflies produce their alluring ghostly glow by mixing oxygen with special molecules in the body to create a chemical reaction that results in the production of light.

The male big dipper firefly flies with a particular jerky flight pattern that makes its light show look a little bit like the Big Dipper star constellation from which it takes its name.

SIZE: 10–14mm long (adult)

DIET: Insects, earthworms and snails

FOUND IN: Meadows and woodland edges throughout many parts of the USA

HARLEQUIN LADYBIRD
(Harmonia axyridis)

Because this ladybird has a tremendous appetite for aphids, farmers across the world once imported it to help them to protect their crops. Sadly, it spread and began to feed on other animals, including other ladybird species. Today, the harlequin ladybird is one of the world's most invasive insect species.

Like many ladybirds, the harlequin ladybird has spots on its body. In the harlequin ladybird, however, these spots can be very variable. While some individuals have no spots, others can have two, four or even 22 spots.

SIZE: 5.5–8.5mm long

DIET: Aphids and the eggs and larvae of other ladybirds

FOUND IN: Originally from eastern Asia, now known throughout North America, Europe and parts of Africa

INSECTS OF CAMOUFLAGE

The tough exoskeleton of insects, and the variety of colours they can evolve to take, mean that some insects are masters of camouflage. But not all camouflage is for hiding from predators, as some of these examples show.

EUROPEAN MOLE CRICKET
(Gryllotalpa gryllotalpa)

Just like a mole, the European mole cricket has a pair of powerful forelegs that help it to dig through soil. To protect itself and stay hidden within the soil, it is covered in tough, velvet-like hair. Mole crickets are rarely seen, but they are easy to hear.

Like other crickets and grasshoppers, the European mole cricket 'sings' by scraping its legs against its body. The male digs a hole from which its echoing calls reverberate. Its call is so loud that nearby soil can be seen to shake.

SIZE: Up to 46 mm long

DIET: Insect larvae and earthworms

FOUND IN: Damp soils in floodplains and wetland edges throughout Europe

CHAN'S MEGASTICK
(Phobaeticus chani)

With its legs outstretched, this recently discovered stick insect is longer than a medium-sized dog. Some scientists think it may have remained undiscovered for so long because it hides so well in the upper canopy of Borneo's rainforests.

Because insects lack lungs and a powerful heart which can pump large volumes of oxygen around the body, today's species are restricted in how big they can get. It is likely that Chan's megastick is about as big as any insect can be in the world's atmosphere as it is today.

SIZE: Up to 60cm with legs outstretched

DIET: Leaves of trees and shrubs

FOUND IN: Remote rainforests in Borneo

WALKING FLOWER MANTIS
(Hymenopus coronatus)

Hiding among orchid flowers, the walking flower mantis waits. Its colour exactly mimics that of the flower and its legs look like orchid petals. It gently rocks back and forth as if it were a flower blowing in the breeze. This insect is supremely camouflaged, but not for the reason you might suspect. It is not hiding from predators: it is an ambush predator itself. When an insect approaches, assuming it to be a flower, the walking flower mantis strikes out with dagger-like legs. The prey is quickly pulled apart and eaten.

There are 2,400 mantis species. Most are ambush predators, although some chase prey on the ground.

With nearly all mantises the females are large, and frequently attack – and even eat – the males. Scientists are still trying to work out what the mantises gain from this bizarre behaviour.

SIZE: Up to 6cm long (females), up to 3cm long (males)

DIET: Insects including crickets, flies, beetles and bees

FOUND IN: Throughout the rainforests of Southeast Asia

BUTTERFLIES AND MOTHS

Together, butterflies and moths are known as lepidopterans. These insects are covered in tiny scales and possess drinking-straw-like mouthparts for sucking up sugary fluids. In all, this group makes up 10 per cent of all life on Earth. Caddisflies are close cousins of lepidopterans. Though hard to spot, caddisflies are no less beautiful.

GROTE'S TIGER MOTH
(Bertholdia trigona)

The Grote's tiger moth has a secret weapon which it uses against its mortal enemy, the bat. When fleeing, it produces special clicks (up to 4,500 each second) which block bats' ability to echolocate. The bats fly straight past this moth, unaware that they have been tricked.

Like most other moths, the Grote's tiger moth is nocturnal. When at rest, it holds its wings firmly along the upper surface of its body. This is different from butterflies, which hold their wings above the body when resting.

SIZE: 2cm long

FOUND IN: Throughout the southwestern states of the USA

DIET: Caterpillars feed on moss and algae, adults feed on flowering plants

LAND CADDISFLY
(Enoicyla pusilla)

Most caddisflies have a larval stage that lives underwater, but the land caddisfly provides another habitat for its young. Its larvae hide among leaf litter under oak trees, living in special protective cases made of sand grains. Unlike other caddisflies, the adult female of this species doesn't fly. She prefers instead to let males seek her out among the leaves.

Caddisflies are very ancient members of the insect group. Fossils of their larval cases suggest that they were around many millions of years before the dinosaurs.

SIZE: 5mm long

FOUND IN: Oak woodlands throughout Europe

DIET: Larvae feed on dead oak leaves

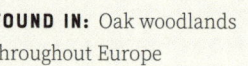

MONARCH BUTTERFLY
(Danaus plexippus)

The monarch butterfly is one of the world's most epic travellers. Each year, hundreds of thousands of them make their journey north from Mexico, heading towards the United States and Canada. As it migrates, a monarch butterfly lays eggs on food plants; these hatch into caterpillars which pupate and turn into a new generation of adults that will continue the journey northwards. When summer ends, the monarchs go back south to overwintering hideouts nearer the equator, where they see out the colder months.

Scientists are still trying to work out how the monarch butterfly manages such an incredible journey. It may be that it is guided by special chemicals left by previous generations on trees or that it uses an internal 'sun-clock' that helps it to track and monitor the changing of seasons.

SIZE: 8.9–10.2cm wingspan

DIET: Caterpillars feed on milkweed, adults drink nectar from a variety of flowering plants

FOUND IN: From southern Canada to northern South America, as well as islands in the Pacific and Atlantic Oceans

FLIES AND THEIR RELATIVES

With a million species or more, flies are an incredibly important part of the insect family tree. They are known for having one pair of wings, rather than two. Some fly relatives, such as the scorpionfly, are not true flies but representatives from an early age of insects that lived alongside the dinosaurs.

AMERICAN HOVERFLY
(Eupeodes americanus)

Like many hoverflies, the American hoverfly mimics a wasp or bee to make predators think twice before attacking it. It hovers by beating its wings more than 100 times per second. When it discovers a flower, it drinks precious nectar with a set of mouthparts much like a tongue.

Because hoverflies go from flower to flower in their search for food, they carry lots of pollen between plants, helping them to reproduce. Without flies like these, many of the world's flowering plants would die out.

SIZE: 9–12mm long

FOUND IN: Flower-filled meadows in North America

DIET: Nectar from flowers

VINEGAR FLY
(Drosophila melanogaster)

Because it lays lots of eggs and can be kept easily in a laboratory, the vinegar fly is helping scientists discover many things about how animals work. By studying this fly, scientists have come to understand many important principles including inheritance, variation, genetics and the workings of our DNA.

Like many insects, the vinegar fly does not live for long. Depending on its surroundings, it might die of old age just 21 days after hatching from its egg.

SIZE: 2.5mm long

FOUND IN: Throughout Africa, Europe and Asia and has been transferred by humans to all continents except Antarctica

DIET: Rotting plants and fruits

VIOLET BLACK-LEGGED ROBBERFLY
(Dioctria atricapilla)

The violet black-legged robberfly is rarely off duty. From a special perch it scans overhead flying insects, looking for the most nutritious passers-by. When it spots suitable prey, it thrusts itself skywards with its powerful wings, grabbing prey in mid-flight before pulling it to the ground to drain it of its bodily fluids using its sucking mouthparts.

Like other robberflies, the violet black-legged robberfly has a dense moustache of thick hairs on its face. These hairs protect the fly from its prey's desperate kicks.

SIZE: 9–12mm long
DIET: Flying insects
FOUND IN: Grasslands throughout Europe and Asia

COMMON SCORPIONFLY
(Panorpa communis)

With fierce jaws and a swollen pointed tail, the male common scorpionfly looks deadly yet is completely harmless to humans. It uses its large mouthparts to pull apart dead insects, which it tugs from spiderwebs when the spiders aren't looking. Its tail is a special grabber for holding onto females.

Though the common scorpionfly looks like a fly, it belongs to a different insect group – the Mecoptera. These prehistoric flying insects are more closely related to fleas than they are to true flies.

SIZE: 30mm long
DIET: Dead insects
FOUND IN: Hedgerows throughout Europe and northern Asia

DRAGONFLIES AND DAMSELFLIES

Dragonflies and damselflies are like the falcons of the insect world: fast, fearsome flying predators. They hunt other flying insects on two pairs of powerful wings and have strong biting mouthparts to pull apart their prey. Their young are reared in water where they play an important part in wetland food chains.

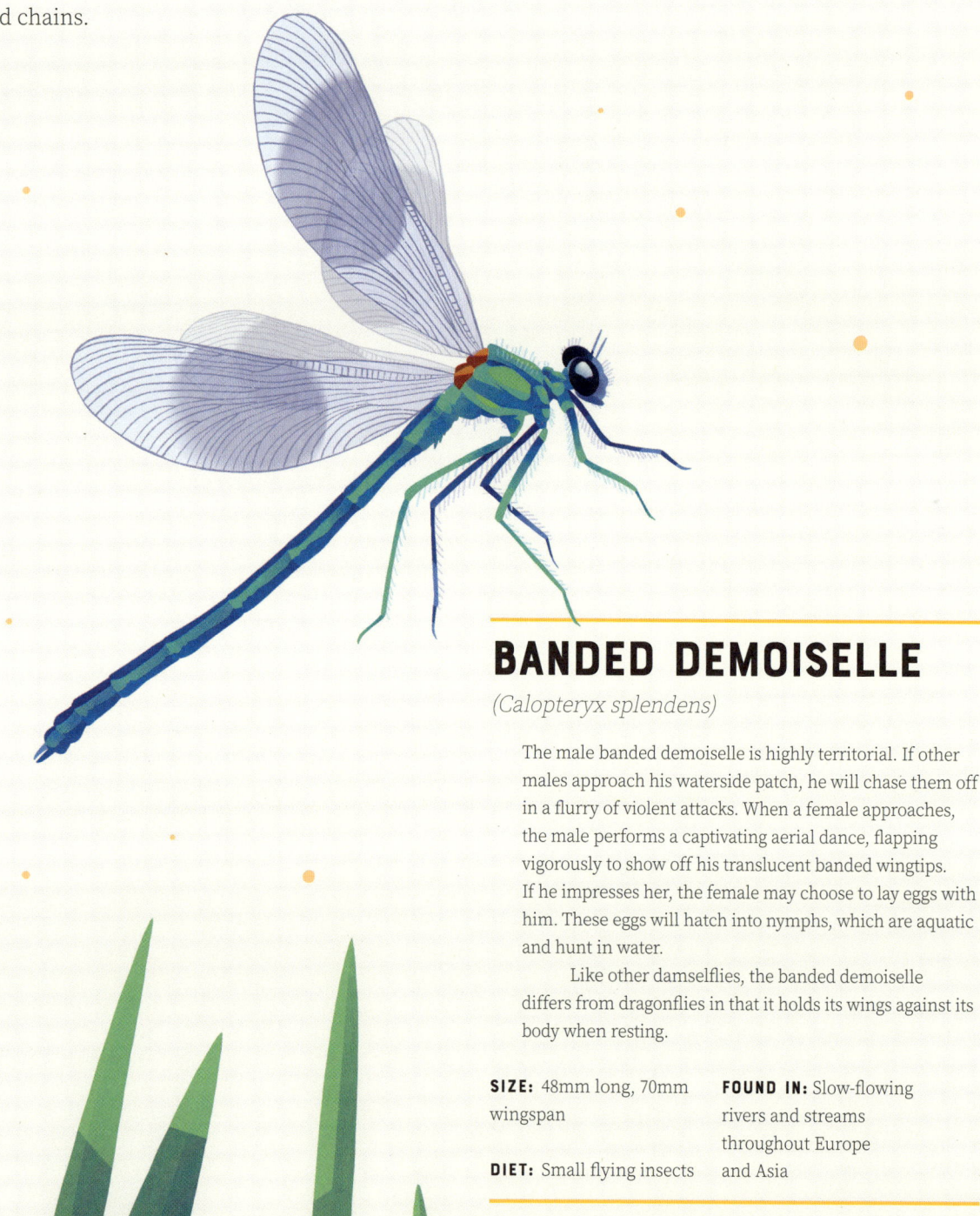

BANDED DEMOISELLE
(Calopteryx splendens)

The male banded demoiselle is highly territorial. If other males approach his waterside patch, he will chase them off in a flurry of violent attacks. When a female approaches, the male performs a captivating aerial dance, flapping vigorously to show off his translucent banded wingtips. If he impresses her, the female may choose to lay eggs with him. These eggs will hatch into nymphs, which are aquatic and hunt in water.

Like other damselflies, the banded demoiselle differs from dragonflies in that it holds its wings against its body when resting.

SIZE: 48mm long, 70mm wingspan

DIET: Small flying insects

FOUND IN: Slow-flowing rivers and streams throughout Europe and Asia

PARASITIC INSECTS

Being small and adaptable, many insects have taken to a parasitic way of life, which means they steal nutrition directly from other animals and plants. Many blood-drinking parasites have mouthparts which work a little bit like a sharp drinking straw.

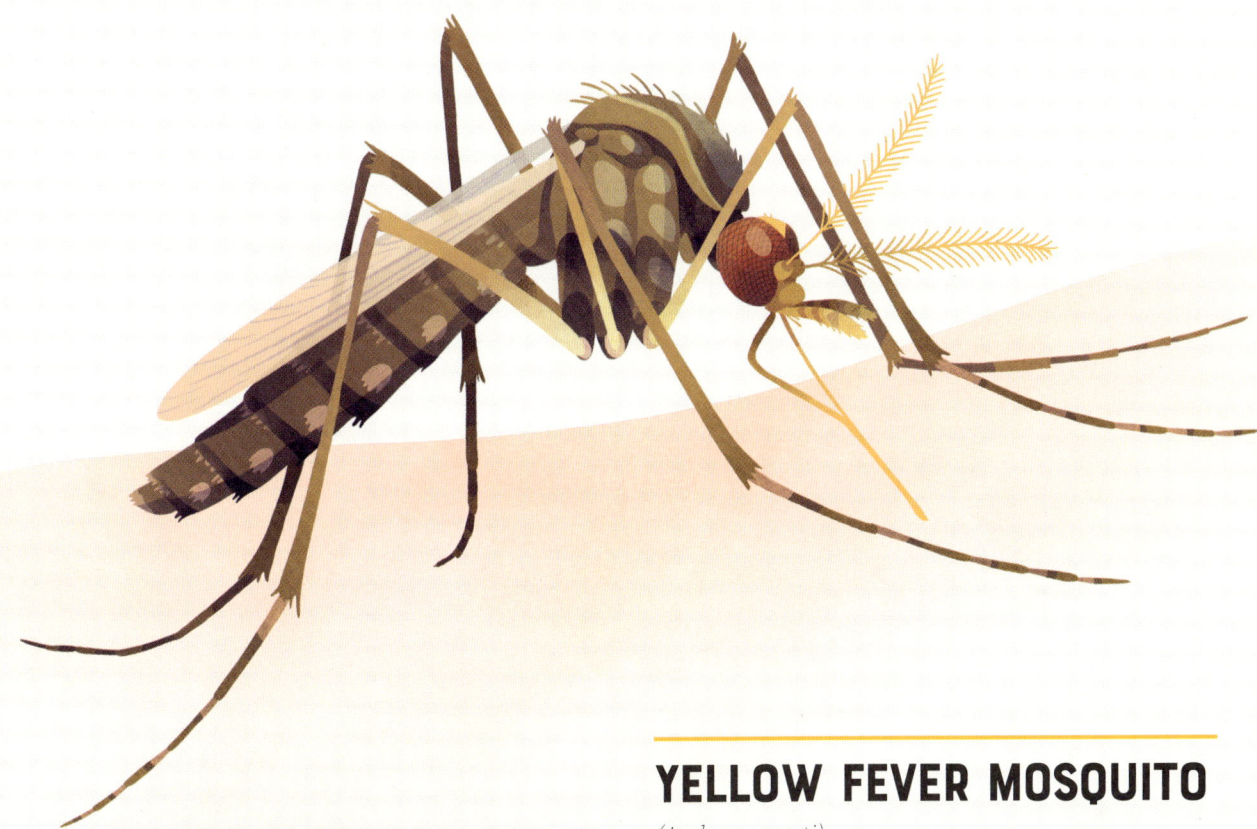

YELLOW FEVER MOSQUITO
(Aedes aegypti)

Few creatures on Earth have made more impact on humans than the yellow fever mosquito. By drinking blood from humans, this species of mosquito can accidentally transmit a host of diseases from person to person, causing death and suffering in the process. Thankfully, more and more of these diseases can be treated with medicine.

As with all mosquitoes, it is only the female yellow fever mosquito which feeds on blood. Males buzz from flower to flower, feeding on sugar-rich nectar. In this way, not all mosquitoes are pests. Many can be important pollinators of flowering crops.

SIZE: 1.8–3.2mm wingspan

DIET: Females feed on blood, males drink nectar

FOUND IN: Originally from the wetlands of Africa, this species now occurs across many tropical and subtropical regions

MYRIAPODS

The name 'myriapod' refers to the myriad (numerous) legs that all members of this arthropod group possess. Mostly, millipedes have more legs than centipedes, sometimes more than 1,000. Together, scientists have described more than 16,000 myriapod species.

AMAZONIAN GIANT CENTIPEDE
(Scolopendra gigantea)

The Amazonian giant centipede regularly hunts amphibians, small reptiles and small rodents. Some individuals have even been known to hunt bats, which they manage by dangling down from the roof of a cave, grabbing the bats as they fly past.

Centipedes differ from all other arthropods because they inject their venom not with fangs but with a set of modified forelegs called forcipules. As with spiders, their venom acts to paralyse prey, limiting an animal's chance of escape when it's attacked.

SIZE: Up to 30cm long

DIET: Large spiders, insects, amphibians, small reptiles and small mammals

FOUND IN: Tropical and dry forests throughout northern parts of South America

WINGLESS ARTHROPODS

Some arthropods provide scientists with knowledge about the earliest invertebrates and what they were like. Silverfish teach us about the earliest insects whilst snow fleas help us understand more about springtails, an early rival to prehistoric insects.

SILVERFISH
(Lepisma saccharina)

The silverfish gets its common name from its shiny, scalelike exoskeleton and the zigzagging way it moves along the ground. At night, it scurries along the floors of houses looking for leftover food scraps rich in sugar. If food becomes scarce, a silverfish can survive quite happily for a year or more on water alone. In really extreme circumstances, some have been known to eat book bindings and wallpaper glue.

Unlike most invertebrates, silverfish are incredibly long-lived. Some individuals can live up to eight years.

SIZE: 13–25mm long

FOUND IN: Throughout the world in human habitations

DIET: Sugar-rich foods

SNOW FLEA
(Hypogastrura nivicola)

The snow flea is actually a type of springtail, a group of tiny arthropods that differ from insects because they have no external mouthparts. During the winter, the snow flea is often seen walking on the surface of snow. It survives in frozen environments by pumping its body with sugar-like proteins which stop its cells from freezing solid.

Springtails are one of the most numerous arthropod groups on Earth. In some parts of the world, a single square metre of soil can be home to 100,000 springtails.

SIZE: 1–2mm long

FOUND IN: Snowfields in North America.

DIET: Fungi, plants and animal remains

WHAT IS A FISH?

Today, in a thousand ways, one world-changing group of animals has exploited the 70 per cent of the world's surface that is covered in water. We call these bony – and sometimes not-so bony – animals fish.

Fish are vertebrates (animals with a backbone or spine). They are found in nearly every aquatic habitat on Earth. They have colonised the highest mountain streams, coral reefs, frozen ponds and even puddles. Some, like the lungfish, can live for short periods in wet mud. Others, like the ornamented flying fish, leap clear of the waves and make the air a temporary safe haven from predators below. In the form of eels and chimaeras, fish exist in many of the deepest oceans. In all, more than 35,000 fish species are known and many more are awaiting discovery by intrepid scientists.

Fish are a vital part of ocean and freshwater food chains. Their tiny larvae are a key ingredient in the vast shoals of plankton that whales and manta rays consume. The middle rungs of food chains include enormous shoals of herring and tuna. Then there are the top predators, including fish such as the famous great white shark.

Throughout their 500 million-year history, fish have survived many mass extinction events, including the meteorite that killed the dinosaurs. Today, however, they are a vital food source for humans. Each year, more than 154 million tonnes of fish are harvested from lakes, rivers and oceans across the world, including 100 million or more sharks.

This global rise in fishing industries has caused many fish stocks to decline, forcing many fish species to the very edge of extinction. Scientists are currently investigating how fish might fare against this, their newest threat.

A typical ocean food chain

FISH CHARACTERISTICS

SWIM BLADDER
Many fish have a swim bladder, a special bag inside the body which can be temporarily filled with or emptied of gases. This allows the fish to move upwards or downwards in the water. Many fish also use their swim bladder as an echo chamber to create sounds used for communicating with one another.

COLD-BLOODEDNESS
Unlike mammals and birds, most fish cannot generate body heat on their own, which means they are cold-blooded (ectothermic). However, some fish, such as tuna and the great white shark, have evolved ways to recycle a small amount of the body heat produced in their muscles, allowing them to move into cooler waters where other fish may struggle.

SCALES
Most fish are covered with scales. Scales offer some protection from predators and parasites and can also help make fish smooth and streamlined in the water, ensuring that they use as little energy as possible whilst travelling. Many fish have scales that reflect light to scare off rivals. Other fish use scales to assist with camouflage, either for ambushing prey or to hide from predators.

GILLS
The gills are where oxygen in the water can pass directly into the fish's bloodstream. Most fish have gills on the side of their head, behind the jaws. But some fish can use a modified swim bladder to take oxygen directly from the air, in the same way that bony land animals do.

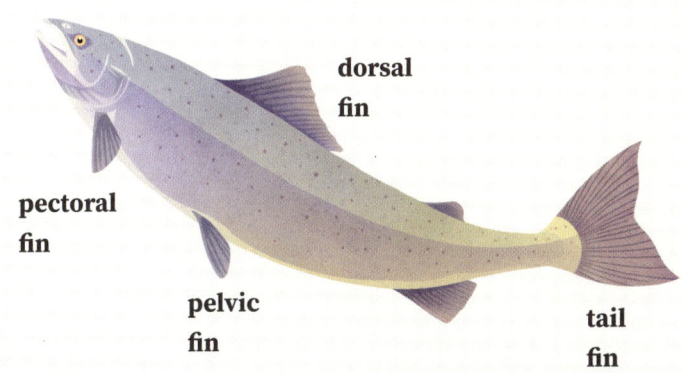

pectoral fin · dorsal fin · pelvic fin · tail fin

FINS
Most fish have two pairs of fins (pelvic and pectoral fins) on the lower side of their body, a pair of fins (dorsal fins) along the upper side and a tail fin which, for most fish, provides thrust. Fins aid in streamlining the body, as well as helping with steering and propulsion.

SHARKS

Sharks include some of the most ancient and the most predatory fish. They are known for their skeletons made of cartilage (a firm but flexible structure different from bone), their thick scales and their rows of self-replacing teeth. In all, there are more than 500 shark species.

SHORTFIN MAKO
(Isurus oxyrinchus)

The shortfin mako is a supercharged shark. Whilst chasing prey, it regularly reaches speeds of 74 kilometres per hour and it can jump more than nine metres out of the water. To power such an energetic lifestyle requires lots of food – it must consume more than 3 per cent of its body weight each day or it will starve.

Unlike many sharks, the shortfin mako does not hunt by sensing electricity given off by struggling prey. Instead, it hunts using ultrasensitive eyes and a keen sense of smell.

SIZE: Up to 4.45m long

DIET: Mostly squid and fish including mackerel, tuna and swordfish

FOUND IN: Temperate and tropical seas around the world

TIGER SHARK
(Galeocerdo cuvier)

With its large and powerful jaws, the tiger shark has a reputation for eating almost anything. As well as feeding upon classic shark prey like fish and squid, it has also been known to hunt sea snakes, birds, crabs and, sometimes, other sharks. And that's just the start. Scientists looking at the stomach contents of dead tiger sharks once discovered the remains of even stranger animals, including rats, horses and fruit bats. Quite how some tiger sharks are able to catch creatures like these is still a mystery.

SIZE: Up to 5m long

DIET: A wide variety of large sea creatures

FOUND IN: Coastal waters of tropical and subtropical countries

GREAT WHITE SHARK
(Carcharodon carcharias)

The great white shark is one of the most feared fish in the ocean, but there is more to this mysterious shark than meets the eye. First, it isn't always in the mood for killing. Far from it. It can go without feeding for long periods by surviving on stored fats and oils in its liver. This means it can travel long distances through barren seas on its way to the richest feeding grounds.

Scientists are still trying to work out where male and female great white sharks meet to mate. Perhaps it is when they gather around dead whales out at sea, which they devour in large and excitable groups.

Unlike most sharks, the great white shark regularly hunts in cold water. To survive in cooler waters, it has a large insulated body with special veins and arteries that recycle body heat.

SIZE: Up to 6m long

DIET: Fish, marine mammals, sea turtles and seabirds; it also scavenges dead whales

FOUND IN: Coastal and offshore waters in many parts of the world, particularly near seal colonies

GREAT HAMMERHEAD
(Sphyrna mokarran)

The great hammerhead has a head like a metal-detector. Its wide face is covered with special cells that detect the electricity produced by prey hiding under the sand, especially stingrays. When a stingray dashes away upon being discovered, the great hammerhead strategically bites off one of its fins to stop it escaping, before tearing it apart with vicious shakes of the head.

The great hammerhead is one of many sharks harvested and killed for its fins to make shark-fin soup. So few remain that it is now critically endangered.

SIZE: Up to 6.1m long

DIET: A wide range of invertebrates and fish, including rays, skates and even other sharks

FOUND IN: Coral reefs and coastal lagoons in tropical regions

WHALE SHARK
(Rhincodon typus)

With a giant mouth and 300 rows of tiny teeth, the whale shark could easily be confused for a man-eater. Yet this shark does not use its teeth for feeding. Instead it uses sieves in its gills to catch millions of tiny planktonic animals in the water, which it swallows greedily.

The whale shark is easily the largest shark in the ocean, dwarfing the great white shark and the tiger shark. Incredibly, scientists are yet to discover where in the ocean its nursery grounds might be.

SIZE: Up to 17m long

DIET: Plankton: baby squid and fish, tiny shrimps, fish eggs and crab larvae

FOUND IN: Open waters in tropical oceans across the world

SPOTTED WOBBEGONG
(Orectolobus maculatus)

The spotted wobbegong gets its name from an Aboriginal Australian word for 'shaggy beard'. This is because its mouth is surrounded by whiskery frills which look like the seaweeds and coral outcrops within which it hides. This beard, along with spots and sandy markings on its back, gives the spotted wobbegong a cloak of invisibility. If a fish should accidentally stray too close, the wobbegong rises up in the blink of an eye, its mouth held wide open to swallow its prey whole.

SIZE: Up to 3m long

DIET: Fish

FOUND IN: Warm waters along the west coast of Australia

DWARF LANTERNSHARK
(Etmopterus perryi)

The dwarf lanternshark is such a small shark that it could fit comfortably in a human hand. It is covered in cells that produce light, which suggests that it can give off a ghostly glow in the dark ocean depths. Many of these light-producing cells are found on the underside of its body. This makes it harder for hungry predators to see it from below.

Very few specimens of this deep-sea shark have ever been found, so scientists are working hard to learn more about it.

SIZE: Up to 20cm long

DIET: Small fish and tiny squid

FOUND IN: Deep oceans off the coasts of Venezuela and Colombia

RAYS

There are more than 600 species of rays. Like their shark cousins, they are known for their soft, cartilaginous skeletons. Rays' pectoral fins are greatly elongated and fused to the head. All rays move by flapping or undulating these long, winglike fins.

BLUESPOTTED RIBBONTAIL RAY
(Taeniura lymma)

At night, the bluespotted ribbontail ray searches for prey. It flaps towards sandy patches of sea floor where it forms small shoals with other rays. These shoals move carefully over the sand scanning it for invertebrates. The bluespotted ribbontail ray has tough, crushing jaws capable of smashing mollusc shells or the armoured exoskeletons of crabs and lobsters.

This ray is a stingray: it has a venomous barb on the rear of its body. When threatened, it can use this barb against an attacker, delivering a poison which causes painful muscle cramps. Stings on humans are rare.

SIZE: 70cm long including tail

FOUND IN: Coral reefs in the tropical Indian Ocean and western Pacific Ocean

DIET: Molluscs, worms, shrimps, crabs and fish

ATLANTIC TORPEDO
(Tetronarce nobiliana)

When threatened, the Atlantic torpedo can deliver a mighty shock. Using special electricity-generating organs in its body, it can produce sudden bursts of up to 220 volts of electricity, easily enough to knock a human unconscious. Because of this electrical weaponry, this ray has few known predators.

Like other cartilaginous fish, the Atlantic torpedo has 'distensible' jaws that can stretch out from its face. This allows it to eat very large prey.

SIZE: Up to 1.8m long

FOUND IN: Cooler waters on both sides of the Atlantic Ocean

DIET: Bony fishes and some crustaceans

GIANT OCEANIC MANTA RAY

(Mobula birostris)

The giant oceanic manta ray is the largest ray in the world. Its enormous wings and streamlined shape allow it to cruise through the ocean almost effortlessly as it journeys towards the finest feeding grounds.

When it finds an area rich in swimming shrimp, planktonic crabs and krill, this huge ray herds these tiny prey into dense clouds which it then rushes towards with its mouth open wide. As well as filter-feeding like this, scientists have recently discovered that the giant oceanic manta ray can also hunt and devour deep-sea fish.

SIZE: 7m across

DIET: Shrimps, plankton and small fish

FOUND IN: Tropical and temperate waters across the world

GHOST SHARKS

'Ghost shark' is the common name given to a fish group called the chimaeras. Chimaeras are closely related to sharks but their skins lack the tough armour of sharks and, unlike sharks' teeth, their teeth are not replaceable. Most chimaeras live in the deep sea.

PACIFIC SPOOKFISH
(Rhinochimaera pacifica)

The Pacific spookfish has an incredibly long nose that it uses to sense small fish swimming nearby in the dark waters of the deep sea. To protect itself from deep-sea predators, it has a venomous spine on its dorsal fin, which encourages animals to think twice before attacking it.

Many sharks have teeth that they can replace again and again if lost. But the Pacific spookfish, like other chimaeras, does not have teeth like this. Instead, it crushes prey using three pairs of large, grinding teeth.

SIZE: Up to 130cm long

DIET: Small fish

FOUND IN: Deep areas of water throughout the Pacific Ocean

SPOTTED RATFISH
(Hydrolagus colliei)

Unlike other ghost sharks, the spotted ratfish occasionally ventures into shallow waters. Here it searches for hard-shelled prey such as clams and crabs. As with other chimaeras, its upper jaw is locked into its skull. This gives it an extra-tough grip on its armoured prey. Its muscular jaws and sharp front teeth allow it to bite with more force than any other known chimaera. It gets its name from its unusually long ratlike tail.

SIZE: Up to 1m long

DIET: Molluscs and crustaceans

FOUND IN: Throughout northeastern regions of the Pacific Ocean

JAWLESS FISH

Today's jawless fish hint at what the very first fish may have been like, more than 400 million years ago. These primitive fish have simple light-gathering eyes and quite a basic gut that lacks a true stomach. This once-blooming part of the fish family tree has just 120 representatives living today.

EUROPEAN RIVER LAMPREY
(Lampetra fluviatilis)

The European river lamprey is simple but ferocious. It seeks out larger fish which it attaches to with a sucker-like face. When it is attached, its razor-sharp teeth scrape against the larger fish's scales, opening up a wound from which the lamprey feeds.

The European river lamprey has a complicated life cycle which involves adults travelling from the ocean upstream into rivers and streams to lay eggs. The enormous effort involved in this migration is so great that once eggs have been laid, all of the adults die.

SIZE: 25–40cm long

FOUND IN: Coastal waters across Europe

DIET: A parasite of larger fish

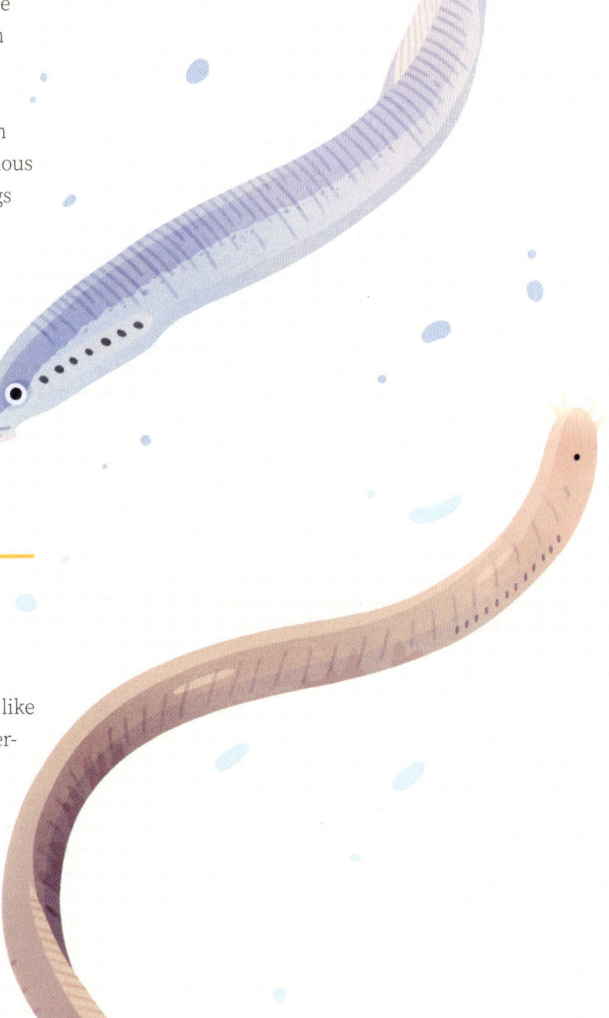

ATLANTIC HAGFISH
(Myxine glutinosa)

With no obvious eyes or jaws, the Atlantic hagfish looks like a giant swimming zombie worm. With its sensitive finger-like whiskers it detects the smells of decomposing animals in the water, which it slides into to eat from the inside out.

Hagfish are famous for their slime-making abilities. When caught in the jaws of a predatory fish, a hagfish can produce such a volume of slime that it can block up its attacker's gills. For this reason, few predators willingly seek out the Atlantic hagfish.

SIZE: Up to 75cm long

FOUND IN: Sea floors across the eastern coasts of the Atlantic Ocean

DIET: Dead fish and other large animal carcasses

BICHIRS

Unlike most fish, members of the group known as bichirs have not just two but anything from seven to eighteen dorsal fins. Bichirs are covered in thick bonelike scales and they have paired lungs which allow them to breathe air.

REEDFISH
(Erpetoichthys calabaricus)

Like other bichirs, the reedfish has a pair of lungs in addition to its gills. This means that, should the water get too low in oxygen, it can come up to the water surface to take in a mouthful of air.

This air-breathing adaptation has allowed the reedfish to take on an impressive new hunting style. If food in the water is scarce, it can use its powerful pectoral fins and strong tail to slither out of the water to search for land-living invertebrates.

SIZE: Up to 37cm long

DIET: Worms, crustaceans and insects

FOUND IN: Slow-moving freshwaters across many parts of central Africa

GUINEAN BICHIR
(Polypterus ansorgii)

The Guinean bichir is a shy and graceful freshwater predator that hunts during the twilight hours.

Like other bichirs, it has lots of tiny dorsal fins (called finlets) that run along its back. Its pectoral fins are fleshy and strong to help it to move along the lake floor. Thick bony scales cover its body to protect it from predators such as crocodiles and eagles.

Encounters with this species are so uncommon that scientists get very excited when it is spotted in the wild.

SIZE: Up to 28cm long

DIET: Crustaceans, insects and other small invertebrates

FOUND IN: Rivers and lakes across western Africa

BOWFINS AND GAR

These two closely related groups of fish have cartilaginous skeletons covered in bony armour. This means they have skeletons that resemble those of both sharks and bony fish. Bowfins and gar live in freshwater almost all of the time, though some can tolerate saltwater for short periods.

ALLIGATOR GAR
(Atractosteus spatula)

The alligator gar is one of the largest freshwater fish in North America. It swims silently and stealthily before striking sideways at unwary fish using its long jaws filled with double rows of spiky teeth.

Like its bowfin cousins, the alligator gar has a special swim bladder that can function as a primitive air-breathing lung. In the hot summer when oxygen levels in the water drop, it is often heard taking a deep gasping breath at the water surface.

SIZE: Up to 2.5m long

DIET: Fish and occasionally waterfowl and small mammals

FOUND IN: Lakes and reservoirs throughout the southern USA

BOWFIN
(Amia calva)

The bowfin is known for its impressive jaws, capable of biting and killing prey within a fraction of a second. Like its close cousins, the gars, a bowfin can also take oxygen directly from the air. Unlike gars, however, a bowfin's scales are more rounded and less bony.

During the age of dinosaurs, members of this family were found throughout the world's oceans and freshwaters. Today, there are very few. Scientists are still debating the cause of their global decline.

SIZE: 50cm long (average)

DIET: Fish and large crustaceans

FOUND IN: Lowland rivers, lakes and swamps throughout North America

FISH WITH STRANGE LARVAE

Some fish, including eels, produce tiny babies with flattened bodies that are almost totally transparent. These fish larvae are known as leptocephali. Leptocephali drift for many months feeding on microscopic plankton. Scientists are still trying to discover more about the strange life stage seen in this group of fish.

ATLANTIC TARPON
(Megalops atlanticus)

Tarpon are famous for their upward-pointing jaws and a tough bony plate on the lower jaw capable of crushing crustaceans into tiny pieces. Their bright, shiny scales give this species the nickname, the silver-king.

When fully grown, a female Atlantic tarpon can produce as many as 12 million eggs, most of which hatch into tiny see-through larvae, each smaller than an ant in size. Nearly all of these baby fish hide in clouds of plankton where they will become prey to whales, manta rays and filter-feeding sharks.

SIZE: Up to 2.5m in length

DIET: Small fish and crustaceans

FOUND IN: Tropical and sub-tropical regions of the Atlantic Ocean

BONEFISH
(Albula vulpes)

The bonefish can change the chemistry of its cells so that its body can live in both salty and not-so-salty water. Each day it rides the tides, migrating from tropical seas into brackish lagoons and mudflats to find food.

The bonefish uses its paddle-like face to dig into sand for buried invertebrates. Occasionally it will cosy up to stingrays hoping to hoover up prey escaping from the stingrays' crushing jaws. In shallower waters, the bonefish digs so vigorously that its swishing tail can stick out above the water surface.

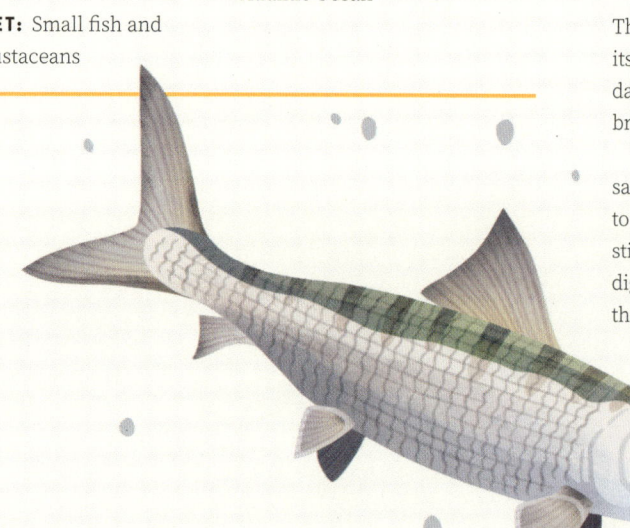

SIZE: Up to 104cm long

DIET: Worms, fish, crustaceans and molluscs

FOUND IN: Atlantic waters throughout Central America and the Caribbean

GIANT MORAY EEL
(Gymnothorax javanicus)

The giant moray eel's jaws are not adapted to suck food into its mouth when they open, as most fishes' jaws do. So it must depend on another trick to grasp its prey. At the back of its throat it has a second set of tooth-laden jaws, called 'pharyngeal jaws'. These nightmarish jaws spring out from between the main jaws when the mouth opens, before snapping back into place. The pharyngeal jaws help the giant moray eel to pull prey quickly down its throat.

This fish is such a skilled predator that it regularly eats poisonous animals. Thankfully for the giant moray eel, this poison does not affect it. Instead, it absorbs the poison into its own flesh, an adaptation which puts off most larger predators such as sharks.

SIZE: Up to 3m long

DIET: Fish and crustaceans

FOUND IN: Lagoons and coral reefs throughout the Indian Ocean and western Pacific Ocean

FRESHWATER OMNIVORES

During rainy seasons, the world's rivers, ponds and lakes offer much by way of opportunity for the bony fish that live there. But when lakes and rivers shrink back in the dry months, fish must adapt. To survive hard times, many freshwater bony fish are omnivores that can change their diet to feast on anything going.

BLACK BULLHEAD CATFISH
(Amciurus melas)

A beard of whiskers covers the black bullhead catfish's face. These sensitive feelers, called barbels, allow it to hunt in the depths of night when other fish aren't active. It also has taste buds all over its body, meaning it can use its skin to sense whether there is food around.

The black bullhead catfish can defend itself from bigger fish by thrashing deadly spines on its pectoral and dorsal fins.

SIZE: Up to 60cm long

DIET: Seeds, leaves, decaying plants and a range of invertebrates and fish, both dead and alive

FOUND IN: Lakes and slow-flowing rivers throughout the central USA

GRASS CARP
(Ctenopharyngodon idella)

The grass carp spends most of its life seeking out the most nutritious freshwater plants, which it consumes at an extraordinary rate. Each day, this hungry fish may eat more than its own body weight in leaves. It takes just eight hours for food to travel from its mouth to the end of its gut.

Although mostly herbivorous, the grass carp has keen eyes and a giant mouth filled with cutting teeth. This means that it is also partial to snapping up any aquatic invertebrates that fail to get out of its way.

SIZE: Up to 150cm long

DIET: Mostly plants but also insects, snails and crustaceans

FOUND IN: Native to East Asia but now introduced to many parts of the world

RED-BELLIED PIRANHA
(Pygocentrus nattereri)

Many people fear the razor-sharp teeth of the red-bellied piranha and its apparent appetite for blood. However, the piranha's reputation is far worse than its bite. In reality, this fish is an omnivore that spends much of its time foraging in the water for fruits, leaves and small animals, especially dead animals.

Often, the red-bellied piranha gathers together with other piranhas in great shoals. Scientists used to think this shoaling behaviour was because the piranhas were pack-hunters. Today, most scientists think that piranhas shoal like this to keep safe from predators such as dolphins, water birds and caiman crocodiles.

Within these shoals, the red-bellied piranha is very social. Each individual piranha communicates with others by making a variety of noises using its swim bladder. Mostly, these sounds are to warn other piranhas in the group to keep their distance. Piranhas, it seems, don't like to share.

SIZE: Up to 50cm long

DIET: Insects, fish, plants and other organic matter

FOUND IN: Throughout the roaring rivers of South America's Amazon basin

FRESHWATER PREDATORS

Many freshwater bony fish have a sharklike reputation. These fearsome predators hunt in a variety of ways, either using their sharp eyes or their inquisitive nature, or using their ability to sit and wait for long periods, primed like a trap ready and waiting to ambush their prey.

NORTHERN PIKE
(Esox lucius)

Many fish are capable of rapidly beating their tails to perform a special fast-dash escape from predators. The northern pike, however, uses this same trick against the fleeing fish. As prey dashes away the pike dashes faster, killing the prey within seconds.

The northern pike is such a monster-like predator that it will even hunt and kill smaller individuals of its own species. For this reason, young stay near the weediest parts of ponds and lakes to make sure they aren't spotted.

SIZE: Up to 150cm long

DIET: Fish, including babies of its own species

FOUND IN: Slow-flowing rivers, lakes and large ponds throughout the northern hemisphere

ATLANTIC SALMON
(Salmo salar)

The Atlantic salmon's migration from the ocean to rivers and streams takes lots of energy. In many species of salmon, the effort is so great that the adults die after laying their eggs. In Atlantic salmon, however, many adults have extra strength and manage to make their way back out to sea for another round.

Within days of hatching, an Atlantic salmon begins to hunt tiny freshwater invertebrates. As it grows and journeys towards the ocean it begins to prey on other animals, including eels and shrimps. Later, as an adult, it will return to the same river to breed.

SIZE: Up to 150cm long

DIET: Invertebrates and small fish

FOUND IN: Rivers and surrounding oceans across Europe and North America

POND SMELT
(Hypomesus olidus)

The pond smelt is a bit like a miniature shark that hunts tiny swimming invertebrates. This little fish uses its large eyes to spot the darting movements of crustaceans, including water fleas, which it gobbles up whole.

The pond smelt is almost totally unique among fish for having miniature teeth on its tongue. Scientists are still examining why it has evolved such a strange adaptation. Perhaps these teeth help it grip onto prey to stop it escaping.

SIZE: Up to 20cm long

FOUND IN: Icy coastlines of the Pacific Ocean

DIET: Tiny freshwater and marine invertebrates

RAINBOW TROUT
(Oncorhynchus mykiss)

With large, sensitive eyes, the rainbow trout scans the surface of the water looking for large beetles and grasshoppers drowning. It also frequently eats the eggs of other fish. In fact, it eats almost anything it can get its jaws on, especially crustaceans.

Because humans enjoy the taste of rainbow trout, this species has been introduced to many countries across the world. Sadly, it has outcompeted many native fish species, forcing some towards extinction.

SIZE: Up to 122cm long

DIET: Invertebrates and small fish

FOUND IN: Originally found in coastal and freshwaters in northern regions of the Pacific Ocean, introduced to many countries

FANTASTIC FINS

Fins have been adapted by fish for a host of jobs. Some fish use their fins to move in unique ways. Other fish use their fins for camouflage. And then there are fish such as the lionfish, whose fins have become a venomous means of defence.

DWARF SEAHORSE
(Hippocampus zosterae)

The dwarf seahorse is the world's slowest-moving fish. When swimming at top speed it moves approximately just 1.5 metres an hour – slower than many snails. Moving at such a slow speed helps it stay undetected by most predators which think of it as nothing more than a insignificant piece of waterweed.

Like other seahorses, the dwarf seahorse has a skeleton covered in bony plates which give it armour-like protection from predators. But this armour makes movement difficult. Unlike most fish, who move their tail from side to side to move forwards, the dwarf seahorse moves by undulating its dorsal fin like a flowing scarf. Its pectoral fins help it steer.

As with other seahorses, it is the male dwarf seahorse who raises the babies. When pregnant, the female gives the male her developing eggs and he keeps them in a special little pouch on his front. The male 'gives birth' to tiny seahorse babies about ten days later.

SIZE: Up to 5cm long

DIET: Tiny crustaceans

FOUND IN: Tropical seagrass beds throughout coastal North America and the Bahamas

RED LIONFISH
(Pterois volitans)

The red lionfish has lots of spiky fins on top of its body, which give it a mane of venomous barbs. When threatened it points its head down to show these fins off, scaring away potential predators.

The red lionfish is such an efficient hunter of fish that it has caused whole ecosystems to change when accidentally released by humans into new areas of the ocean. To stop this species from spreading too quickly, people have been encouraged to eat it as a tasty alternative to more endangered fish species.

SIZE: Up to 47cm long

DIET: Smaller fish

FOUND IN: Native to the Indo-Pacific region but increasingly found along the western coast of the Atlantic Ocean

SEA SPEEDERS

For many species of fish, finding enough prey or regularly escaping from predators requires three things: stamina, speed and streamlining. The fish on these pages are top of the class when it comes to mastering these three principles.

GREAT BARRACUDA
(Sphyraena barracuda)

The great barracuda floats patiently in the water, eyeing up passing prey. Something catches its eye. A pause. Then it strikes. Its tail muscles snap into action, propelling it towards its prey like a bullet. The prey has no time to escape. Within milliseconds it is in the barracuda's jaws.

To hold its food still before swallowing, the great barracuda uses fanglike teeth as well as a secret row of teeth on the roof of its mouth which it uses for grip. These impressive teeth give this giant fish a ghoulish grin.

SIZE: Up to 2m long

DIET: Fish, squid and crustaceans

FOUND IN: Mangroves and deep reefs in warm waters throughout the Pacific, Atlantic and Indian Oceans

BLACK MARLIN
(Istiompax indica)

In water, few other animals can keep pace with the black marlin. This athletic fish occasionally clocks speeds of more than 2 metres per second as it travels through the open ocean looking for food.

The black marlin uses its long and streamlined nose like a sword. When it moves into a shoal of fish it can stun and kill fish by swiping left and right energetically. Though strong, this sword-like nose can sometimes snap off. Incredibly, 'marlin swords' have been lodged into the sides of larger animals, including sharks.

SIZE: Up to 4.65m

DIET: Fish, octopus and squid

FOUND IN: Off the coast of Australia and throughout waters of the Indo-Pacific region

ORNAMENTED FLYING FISH
(Cypselurus callopterus)

When chased by predators, the ornamented flying fish pumps its tail frantically toward the surface. Reaching an explosive speed of 60 kilometres per hour, it blasts from out of the waves before unveiling four winglike fins that allow it to glide more than 200 metres through the air.

When it comes back down to the surface, it can beat its long tail against the surface of the water, travelling like a speedboat along the surface for another 200 metres or more.

SIZE: Up to 30cm long

DIET: Baby fish and other plankton

FOUND IN: Tropical waters off the west coast of Central and South America

PACIFIC BLUEFIN TUNA
(Thunnus orientalis)

Cold water can make some fish sluggish, but the Pacific bluefin tuna has a trick up its sleeve. By recycling the heat generated by its muscles, it can keep its body warmer than the surrounding waters. This primitive version of warm-bloodedness makes it a more active predator than most other fish.

Perhaps more than any other, the Pacific bluefin tuna has a fierce pace that allows it to dash after small fish and squid. In 10-second sprints, it can regularly reach speeds of more than 48 kilometres per hour.

SIZE: Up to 3m long

DIET: Squid and fish, including herring and mackerel

FOUND IN: Throughout the North Pacific

COLOURFUL REEF FISH

Coral reefs account for less than one per cent of the world's oceans yet they provide a home for almost a quarter of all ocean life. To survive in such a busy ecosystem means the fish that live there must stand out to survive.

EMPEROR ANGELFISH
(Pomacanthus imperator)

An adult emperor angelfish has tough jaws for chewing apart spiky sponges found among the coral. To stop sharp bits of food from puncturing its stomach, it covers its food in thick slime after swallowing. Whilst in its juvenile phase, it acts as a cleaner fish, chewing the annoying parasites off larger fish.

This colourful fish is known for its dazzling patterns which make it stand out against the corals. These patterns take almost three years to develop properly.

SIZE: Up to 40cm long

FOUND IN: Tropical reefs of East Africa and the Red Sea

DIET: Sponges and other small invertebrates

BICOLOUR PARROTFISH
(Cetoscarus bicolor)

Without the bicolour parrotfish and its powerful jaws, many coral reefs would become overgrown with green goo. This parrotfish uses its wide jaws to graze algae, which it scrapes off rocks and coral with teeth that never stop growing.

Whilst feeding, it regularly swallows large chunks of rocks and dead coral, which it grinds into a fine white powder that it later squirts out in its poo. These powdery droppings are what make the fine, white sandy beaches that surround many tropical islands.

SIZE: Up to 50cm long

DIET: Mainly algae

FOUND IN: Tropical reefs throughout the Red Sea

FALSE MOORISH IDOL
(Heniochus diphreutes)

For most of its life, the false Moorish idol lives in large shoals looking for surface waters rich in plankton, which it hoovers up with its long jaws. These busy shoals offer protection from predators. To communicate within the shoals, each fish makes special noises to tell others in the group to keep their distance.
This little fish is very inquisitive. Sometimes shoals hundreds-strong will come over to investigate human divers.

SIZE: Up to 21cm long

DIET: Zooplankton in open water

FOUND IN: Sloping reefs throughout the Indo-Pacific Ocean

COMMON CLOWNFISH
(Amphiprion ocellaris)

The common clownfish uses thick slime to protect itself from the stinging tentacles of sea anemones. In return for offering the clownfish protection from predators, the sea anemone uses the clownfish as a form of cleaning service to keep it healthy.
Like many other reef fish, the common clownfish is not strictly male or female but can be both. If a male clownfish becomes the largest in the group, it finds itself turning into a dominant egg-laying female, which comes to rule its anemone colony.

SIZE: Up to 11cm long

DIET: Plankton and algae

FOUND IN: Coral reefs in parts of the Indian and Pacific Oceans

STURGEONS AND PADDLEFISH

Sturgeons and paddlefish are two closely related fish groups with a history going back almost 300 million years. Although they belong to the bony fish group, they actually have skeletons made of cartilage, like sharks, and an incredible spiral-shaped gut which means they can gather every last drop of energy from any meal.

BELUGA STURGEON
(Huso huso)

Occasionally reaching a size that rivals the great white shark, the beluga sturgeon is a monster-like predator of oceans and freshwaters. Some individual beluga sturgeon have even been known to hunt birds and, incredibly, seal pups.

The eggs of the beluga sturgeon (called caviar) are highly prized by humans. Sadly, overfishing has brought this species to the very edge of extinction. Though its family has survived since the age of dinosaurs, the future of this majestic fish is now far from secure.

SIZE: Up to 8m long

DIET: Mostly fish

FOUND IN: Caspian Sea, Black Sea and Adriatic Sea

AMERICAN PADDLEFISH
(Polyodon spathula)

The American paddlefish has a long nose covered in sensitive pits with which it detects electrical fields given off by tiny crustaceans and other planktonic animals, called zooplankton, in the water. It swallows shoals of these plankton using its gills to filter the food.

Paddlefish are some of the oldest surviving representatives of the fish family tree. Fossils suggest that the earliest relatives of modern-day paddlefish were swimming in the oceans almost 300 million years ago.

SIZE: Up to 2.2m long

DIET: Zooplankton

FOUND IN: Found in rivers across 22 states of the USA

LOBE-FINNED FISH

The lobe-finned fish have incredibly strong and flexible fins and tough teeth covered in enamel like our own. This group includes the earliest land-walking fish that began to evolve into the first amphibians and reptiles more than 350 million years ago.

AUSTRALIAN LUNGFISH
(Neoceratodus forsteri)

The Australian lungfish lives in shallow freshwater, often hiding in mud and sand under rocks and logs. If its pond dries up, it can survive for days on end, breathing air using a primitive lung that resembles those of the earliest amphibians.

Lungfish have an impressive fossil history stretching back to an age when lobe-finned fish first took to the land. Incredibly, this means that they are more closely related to bony land animals than they are to other fish such as sharks.

SIZE: Up to 170cm long

DIET: Frogs, tadpoles and fish as well as numerous invertebrates and plants

FOUND IN: Slow-flowing weedy rivers and reservoirs across isolated parts of western Australia

WHAT IS AN AMPHIBIAN?

Amphibians are cold-blooded vertebrates that live on land and in freshwaters across the world. They represent a very prehistoric way of life. Each year, most amphibian species visit water to lay jelly-like eggs. These eggs then hatch into aquatic larvae – often called tadpoles. By living this kind of life cycle, they are like the very first bony land animals that walked the Earth.

About 8,600 amphibian species are alive today and they live in habitats all over the world, from snow-covered grasslands to the driest deserts. Frogs and toads are the most successful part of the amphibian family tree. Together, they make up more than 90 per cent of all amphibians. The rest of this group is made up of the tailed amphibians (often called salamanders and newts) and the wormlike amphibians (caecilians).

Amphibians are one of the most threatened animal groups on Earth: many species have faced extinctions in recent years. This is partly because of a fungal disease which scientists are still investigating. The disease stops amphibian skin from working properly and can cause lots of adult and baby amphibians to die all at the same time. Amphibians also suffer when humans pollute the areas of freshwater that most species depend upon. Because many species breathe through their skin whilst in water, toxic chemicals can easily leak into their bodies affecting their growth and health.

Amphibians have been on Earth for more than 400 million years, but in recent years we have come to realise that their position is far from secure and many are in real danger of dying out.

AMPHIBIAN CHARACTERISTICS

JELLY EGGS
Most amphibians are known for their fishlike larvae which, in the case of frogs and toads, are called tadpoles. These larvae develop from jelly eggs laid in freshwater places, including lakes and ponds. Many tadpoles begin life as algae grazers, but most grow up to become hungry predators or scavengers of dead animals. Some tadpoles, such as those of the paradoxical frog, can even be bigger than the adult frog.

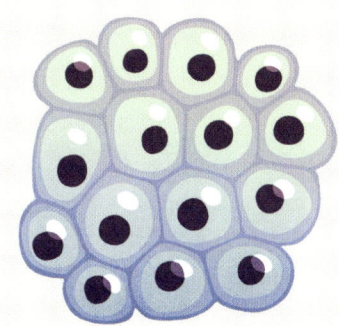

MOIST SKIN
Amphibians don't have waterproof skin like reptiles do (see page 87). This means they can absorb oxygen through their skin, so can breathe underwater. It also means, when on land, they can drink through their skin. Most frogs, for instance, have a special patch of skin on the belly through which they can pull water in from their wet surroundings. The downside of this is that, in dry weather, water can easily evaporate from out of their bodies, dehydrating and killing them within hours.

POISON
Poison is the weapon of choice for many amphibians. Because their skin is thin and lacks waterproofing, they ooze special antimicrobial chemicals, some of which, in the case of the family known as poison arrow frogs, have evolved to become very poisonous. Amphibians let predators know of their poisonous skin using bright colours and vibrant patterns, which predators quickly learn to avoid.

DEFENCE
Some amphibians rely on spiny weapons to keep safe. When threatened by predators, the Iberian ribbed newt, for example, can force its knife-sharp ribs from out of the sides of its body, impaling its victims.

FROGS

Frogs are one of the most abundant animal groups on the planet. There are frogs that climb, frogs that dig and even frogs that can glide. With their large eyes, frogs are watchful and finely tuned predators of invertebrates. Their long legs help them escape predators, though some rely on poisonous skin to keep them extra safe.

RED-EYED TREE FROG
(Agalychnis callidryas)

The red-eyed tree frog is an excellent tree climber, thanks to its slender limbs and sticky pads on its fingers and toes. It is also one of the only bony land animals to communicate using vibrations. For instance, if a male approaches another male during the breeding season, he will violently shake the branch he is on to scare his rival away.

When sleeping in the rainforest canopy, the red-eyed tree frog can cover up its brightest stripes by folding its legs across its body and closing its big, red eyes. Curled up like this, it camouflages perfectly and is harder for predators to spot.

SIZE: Up to 7cm long from nose to tail

FOUND IN: Rainforests and lowlands of Central and South America

DIET: Insects and other invertebrates

AFRICAN BULLFROG
(Pyxicephalus adspersus)

The African bullfrog has a voracious appetite and wide, tooth-filled jaws. Its tiny teeth help it to grip onto animals, such as baby snakes, which it swallows whole. It has even been known to eat members of its own species.

The male African bullfrog is twice the size of the female and can weigh two kilograms. Unlike with most amphibians, the male looks after its babies by digging special canals between puddles to keep them nice and wet.

SIZE: Up to 25cm long

FOUND IN: Ponds and puddles throughout the wetlands of Africa

DIET: Insects, small mammals, reptiles and baby birds

DYEING DART FROG
(Dendrobates tinctorius)

The dyeing dart frog is one of the largest of the poison dart frogs. When eaten by a predator, it produces a vicious cocktail of poisonous chemicals which drip from its skin. The poison causes muscle cramps in its victims and so a predator will quickly spit the frog out and vow never to eat such a colourful frog again.

Indigenous hunters once used the dyeing dart frog's skin secretions to make captive parrots grow specially dyed feathers that were prized amongst local tribes.

SIZE: 5cm long from nose to tail

DIET: Insects and other small invertebrates

FOUND IN: Many rainforests throughout South America

NORTHERN GLASS FROG
(Hyalinobatrachium fleischmanni)

The body of the northern glass frog is so fragile that a single large raindrop can be enough to kill it. Its eerily translucent skin means that through its belly you can see its stomach and other internal organs inside. If it stays still like a statue, this ghostly skin allows it to blend into its surroundings, wherever it is.

Male northern glass frogs are fierce fighters. They have a strange hooklike spike that comes from their spine, which they use as a weapon to defend their territory.

SIZE: Up to 3cm long from nose to tail

DIET: Insects and other small invertebrates

FOUND IN: Fast-flowing streams in parts of Central and South America

TOADS

Toads are a fascinating side-branch of the frog family tree. They often have tough skin and many species produce powerful poisons from special glands behind the eyes. Like many frogs, toads often travel back to the same ponds in which they were tadpoles. Sometimes they have to undertake long and energy-sapping journeys to do this.

COMMON TOAD
(Bufo bufo)

The common toad is a silent but deadly ambush predator. It hides underneath logs and stones waiting patiently for prey to walk too close – and then scoops it up with a powerful snap of its sticky tongue.

Each year when the breeding season begins, the common toad makes a marathon-like journey back to its ancestral pond. Some toads walk more than a kilometre at this time of year, which is equivalent to a human walking 800 kilometres.

SIZE: Up to 15cm long from nose to tail

FOUND IN: Forests and wetlands throughout Europe and Asia

DIET: Insects and other invertebrates

GOLDEN TOAD
(Incilius periglenes)

The golden toad is mysterious and secretive, and hasn't been seen for more than 30 years. Scientists even think it may be totally extinct, although expeditions to search the mountain forests where it once lived continue. If it is extinct, it was likely a victim of a skin disease caused by a mysterious fungus that is behind many amphibian extinctions around the world.

By shuffling its back legs into the soil, the golden toad was able to dig little burrows where it could hide and shelter. Here it waited until night-time, when hunting would begin again.

SIZE: Up to 5cm long

FOUND IN: The cloud forests of Costa Rica

DIET: Small invertebrates

CAECILIANS

The caecilians are a secretive branch of the amphibian family tree. They like to eat earthworms and other digging invertebrates. Overall, there are 200 species and each one lives in stream beds and soils throughout South and Central America, Africa and Southern Asia.

TAITA AFRICAN CAECILIAN
(Boulengerula taitana)

The female Taita African caecilian has a secret. Deep underground in her burrow, she lays eggs that hatch into tiny babies with unusual teeth that are different to her own. The babies use their teeth to tear off chunks of their mother's skin which she produces especially for this purpose.

Like all caecilians, the Taita African caecilian is a soil predator. Using its extra-strong skull, it smashes through soil like a battering ram while it searches for prey. Very little is known of what exactly it eats, though scientists once found lots of remains of termite heads in the stomach of one individual. This suggests it eats more than just worms.

For a long time, scientists have tried to work out where caecilians fit on the family tree of amphibians. It is likely their ancestors split away from frogs and salamanders at some point long before the age of dinosaurs.

SIZE: Up to 35cm long

DIET: Earthworms and other soil invertebrates

FOUND IN: The Taita Hills in southern Kenya

SALAMANDERS

These lizard-like amphibians live mostly throughout the northern hemisphere. Many species live for long periods on land, though some (such as the axolotl) are completely aquatic. Many salamanders advertise that they have poisonous skin with colourful stripes and markings. They also have a remarkable skill – if salamanders lose a leg they can regrow it.

FIRE SALAMANDER
(Salamandra salamandra)

The fire salamander can live for up to 50 years, making it one of the longest-living amphibians. It can reach this age because few predators are interested in eating its poisonous skin, which it advertises with bright patterns.

The poison that fire salamanders produce is called samandarin. If swallowed, this chemical causes muscles to spasm and bouts of high blood pressure in its victims. Such a defence proves more than enough to make most predators think twice before eating one again.

SIZE: Up to 25cm long from head to tail

FOUND IN: Forests and hilly areas of central Europe

DIET: Insects, spiders, earthworms and slugs

GREAT CRESTED NEWT
(Triturus cristatus)

During the breeding season, the male great crested newt becomes like a miniature dragon. Its crest becomes jagged, its belly coloured like fiery flames and it develops a white flash upon its tail. It is ready for love.

Newt eggs differ from frog eggs because, rather than being laid in a big clump, they are laid one by one on the leaves of various pond plants. At first young newts look like tiny fish but within months they begin a predatory way of life on land.

SIZE: Up to 16cm long

FOUND IN: Large ponds without fish across Europe

DIET: Ants, beetles and other insects both in the water and on land

HELLBENDER
(Cryptobranchus alleganiensis)

The hellbender is one of the world's heaviest amphibians. Fiercely territorial, it finds a small patch of rocks and pebbles at the bottom of a stream or river, which it guards with its life.

The hellbender uses a special trick to help it to breathe. It can pull extra oxygen into its body through highly-folded patches of skin. This means that, unlike most amphibians, it rarely needs to come to the surface to breathe.

SIZE: 75cm long from head to tail

FOUND IN: Fast-flowing streams and rivers of the USA

DIET: Crayfish and small fish

AXOLOTL
(Ambystoma mexicanum)

The axolotl is a salamander that never grows up. Instead of developing from a water-living larva into a land animal, it keeps its furry-looking gills even as an adult and stays living in the water. The axolotl locates food using a well-developed sense of smell. When prey gets too close, it can snap its jaws open with such force that water floods into the mouth, sucking the prey in along with it.

The axolotl used to be common but it has become incredibly rare in recent years. This is partly because its freshwater habitats have become polluted.

SIZE: 23cm long or more

FOUND IN: Many lakes near Mexico City

DIET: Worms, insect larvae and small fish

WHAT IS A REPTILE?

Although the dinosaurs suffered extinction, their reptile cousins live on in the form of turtles, snakes, lizards, crocodiles, tuatara and the strange, legless amphisbaenians. In all, there are 10,000 kinds of reptiles alive today and they live in every continent except Antarctica.

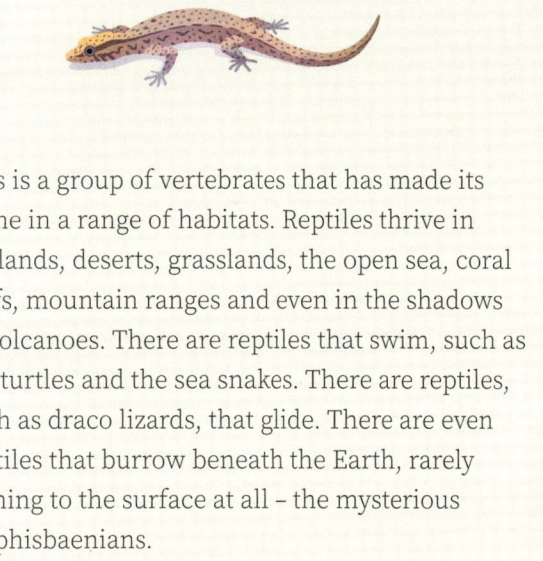

This is a group of vertebrates that has made its home in a range of habitats. Reptiles thrive in wetlands, deserts, grasslands, the open sea, coral reefs, mountain ranges and even in the shadows of volcanoes. There are reptiles that swim, such as the turtles and the sea snakes. There are reptiles, such as draco lizards, that glide. There are even reptiles that burrow beneath the Earth, rarely coming to the surface at all – the mysterious amphisbaenians.

For many years, reptiles were considered less intelligent than birds and mammals. We now know that many reptiles have an impressive brain that allows them to communicate with one another, carefully assessing who is friend and who is foe. Some, like crocodiles, can even use grasses and sticks as a mask of camouflage to help them sneak up on prey. Among the masters of brainpower are those reptiles that have evolved new mechanisms for detecting prey. The most notorious of these are the snakes known as pit vipers, which can 'see' the body heat of their prey.

Reptiles come in a range of sizes. The smallest reptile, at less than the length of a paper clip, is a gecko species called *Sphaerodactylus ariasae*. The largest – a whopping six metres in length – is the saltwater crocodile.

Many of the reptile groups we know of today lived alongside the very earliest dinosaurs, more than 200 million years ago. These prehistoric survivors are likely to include the snakes, turtles, crocodiles and the tuatara. Quite how these groups of reptiles survived the big meteorite impact that finished off their dinosaur cousins is still being debated by scientists.

IDENTIFYING REPTILES

EGGS
Reptiles don't have a tadpole-like life stage like amphibians do. Instead, young reptiles grow in hard-shelled eggs or, sometimes, inside the female's body in a manner very similar to mammals. Most young reptiles are left to fend for themselves but for some reptile species, including crocodiles, the young are cared for and kept safe by at least one parent.

WATERPROOF SKIN
Like mammals, reptiles have waterproof skin. Some have extra bony plates or armour on the skin for protection. Reptiles shed their skins as they grow. In most species this skin comes off in scruffy scraps but in snakes the entire skin is shed, often in one single piece.

COLD-BLOODED
Pound for pound, reptiles can go longer without food than almost any other land vertebrate. This is because most of them do not spend energy keeping their bodies at a particular temperature. If it is cold, their temperature drops and they can slow their bodies right down, finding somewhere to hide until the weather warms up again. It was once thought that all reptiles are cold-blooded, but scientists now know of two groups that are almost certainly warm-blooded: leatherback turtles and dinosaurs.

DEFENCE
Reptiles are most famous for their means of defence. Most notorious are the venomous snakes with their dripping, toxin-laced fangs. Others include the gila monster, with its venomous lower jaws, and the Komodo dragon, which is one of many lizards that comes with a bacteria-filled bite. Though these creatures sound fearsome, most prefer to keep hidden from view and rarely attack humans. Most are masters of camouflage, avoiding humans at all costs.

SEA TURTLES

Sea turtles are master navigators. Some species may travel thousands of kilometres each year, journeying to and from the ancestral breeding grounds that they themselves once hatched from. Sea turtles have an impressive lifespan. Some individuals may reach 80 years old or more.

GREEN SEA TURTLE
(Chelonia mydas)

The green sea turtle has a tough shell that provides protection from most – but not all – predators. When confronted by a shark, particularly a tiger shark, it can wedge itself into gaps in a coral reef to escape the shark's biting jaws.

Unusually for turtles, the green sea turtle is almost totally herbivorous. It chews algae, seaweeds and sea grasses with a special beak that has serrated edges. In fact, it eats so much vegetable matter that its body fat and cartilage turns green, which is how it gets its name.

SIZE: 90cm average

FOUND IN: Coastal areas in more than 140 countries

DIET: Corals, seaweeds, algae, sea grasses

LOGGERHEAD TURTLE
(Caretta caretta)

The loggerhead turtle is the world's largest hard-shelled turtle. It feeds on clams, sea snails, crabs and even sea urchins, crushing them with its powerful jaws. It also has hooked teeth that go all the way down its throat.

The female loggerhead turtle is very protective of her territory. In fact, when threatened she may often fight rival females.

This species of turtle is one of the world's longest-living creatures. It may take 45 years for a female to reach an age where she can lay eggs.

SIZE: Up to 125cm

FOUND IN: Saltwater and estuary habitats around the world

DIET: Bottom-dwelling invertebrates and crustaceans, and sometimes the young of its own species

LEATHERBACK TURTLE
(Dermochelys coriacea)

The leatherback turtle is an intercontinental ocean traveller. Its body is streamlined and its front flippers are long like the wings of an aeroplane. It is the fastest reptile on Earth, occasionally reaching speeds of 35 kilometres per hour in pursuit of jellyfish shoals, which it hoovers up with powerful jaws.

This reptile giant is most noteworthy for its lack of a bony shell. Instead, its carapace (the upper part of the shell) is covered in skin and oily flesh. The leatherback turtle is a lonely survivor of a once-numerous family of sea turtles that stretches back to the age of dinosaurs. It is the only reptile that can generate and recycle its own body heat, almost as if it is warm-blooded. Swimming constantly means its muscles generate heat to keep the body warm. Some individual leatherback turtles have been shown to have core body temperatures that are 18 degrees Celsius higher than the water surrounding them.

SIZE: Up to 2.2m long

DIET: Jellyfish

FOUND IN: All tropical and subtropical waters, even into the Arctic Circle

FRESHWATER AND TERRESTRIAL TURTLES

Many turtles have adapted to a life in freshwater lakes, streams and ponds. There, they quietly feed along the bottom of the water, looking for scraps to scavenge. Some species, like the alligator snapping turtle, are fearsome predators. Many turtle species have even left the water completely: they are called tortoises.

ALLIGATOR SNAPPING TURTLE
(Macrochelys temminckii)

The alligator snapping turtle has a bite so strong it can split a broom handle – and it has an appetite to match. The list of prey that it has been known to eat includes fish, amphibians, snakes, crayfish, worms, water birds, raccoons and even armadillos.

It also has a secret weapon which it sometimes uses to hunt. It has a tongue that looks like a worm, which it dangles out of its mouth like a fishing lure.

SIZE: Shell up to 80cm long

FOUND IN: Freshwaters of the southeastern USA

DIET: Living and dead fish, amphibians and many other freshwater animals

MATA MATA
(Chelus fimbriata)

Perching motionless at the edge of a pond, the shell of the mata mata looks exactly like a floating log and its long neck looks like fallen leaves. Together, these adaptations make it a master of camouflage.

The mata mata is a suction feeder. It uses strange whiskers that dangle from its jaws to feel for pressure waves in the water given off by approaching prey. If the prey gets too close the mata mata lunges open its jaws, sucking the food into its mouth.

SIZE: Carapace up to 45cm long

FOUND IN: Slow-moving stagnant pools throughout South America

DIET: Aquatic invertebrates and fish

YANGTZE GIANT SOFTSHELL TURTLE
(Rafetus swinhoei)

Reaching more than 100 kilograms in weight, the Yangtze giant softshell turtle is the largest freshwater turtle in the world. However, for its size it is incredibly secretive. It spends nearly all of its time swimming in deep waters and rarely comes to the surface to breathe.

With just two individuals surviving today, the Yangtze giant softshell turtle is perhaps the rarest animal on Earth. Scientists are searching for more in the wild. If none can be found, this species will soon face extinction.

SIZE: Up to 1m in total length

FOUND IN: Lakes and rivers of China and Vietnam

DIET: Fish, crabs, snails, frogs, flowers and leaves

GOPHER TORTOISE
(Gopherus polyphemus)

The gopher tortoise is one of the world's most famous habitat engineers. By digging long tunnels in which to sleep, it provides other animals with a free place to shelter from predators, cold weather and forest fires. Approximately 360 different animal species have been known to use its burrows, some of which can be 14 metres long.

The gopher tortoise is a plant scavenger but it will also eat mushrooms and fruits, which it can tear apart with a powerful beak.

SIZE: Carapace up to 30cm long

DIET: More than 300 different plant species

FOUND IN: Deserts of the southeastern USA

CROCODILIANS

Most crocodilians are stalk-and-ambush predators. They wait patiently for unwary prey to get too close before snapping shut their jaws with a bite-force few other predators on Earth can match. These are the closest living cousins of dinosaurs and they are important predators of fish, amphibians, reptiles and occasionally large mammals such as deer, jaguar and even humans.

GHARIAL
(Gavialis gangeticus)

The gharial is a sole survivor of an unusual crocodile family that thrived many millions of years ago. This strange crocodile has long thin jaws lined with 110 sharp, interlocking teeth. This is a jaw shape adapted for snapping at passing fish. Like many crocodiles, the gharial can herd fish into shallow waters where it stuns them by clapping its jaws. It swallows food whole.

Sadly, the gharial has almost been hunted to extinction. 100 years ago there may have been as many as 10,000 gharials, but now only 200 remain.

SIZE: Adults sometimes up to 6m

FOUND IN: Scattered populations throughout the Indian Sub-Continent

DIET: Fish and small crustaceans

AMERICAN ALLIGATOR
(Alligator mississippiensis)

The American alligator often drags its prey into secret underwater chambers where it allows its meal to decompose for a while before eating. This makes it easier for it to tear the prey into bite-sized chunks.

The American alligator is one of two living species of alligator. The other is the Chinese alligator. Alligators are recognisable by their wide U-shaped snouts and the fact that their teeth are more hidden when the jaws are shut.

SIZE: Up to 4.6m long

DIET: Fish, reptiles, amphibians, mammals

FOUND IN: Freshwater including ponds, marshes, wetlands and swamps in the southeastern USA

SPECTACLED CAIMAN
(Caiman crocodilus)

The spectacled caiman is one of six caiman species. It has an unfussy diet that includes insects, crustaceans, fish, snails and, occasionally, wild pigs.

Caimans are close cousins of alligators. Though slightly smaller, caimans have relatively longer and more slender teeth. Their scales are enriched with an extra layer of calcium.

The spectacled caiman is a very protective mother and she shares parental duties with other caimans. Baby caimans are often reared together in special crèches.

SIZE: Up to 2.5m long

DIET: Fish, molluscs, insects, crustaceans, occasionally small mammals

FOUND IN: Freshwaters including swamps, flooded savannahs and mangroves in South and Central America

SALTWATER CROCODILE
(Crocodylus porosus)

The saltwater crocodile is the heaviest of all living reptiles. Because it can regularly kill and eat other top-level predators such as sharks, the saltwater crocodile is classed as an apex predator.

This is a crocodile that can pose a high risk to human life. In fact, as many as 30 attacks on people occur each year, some fatal.

The saltwater crocodile is one of many so-called 'true crocodiles' – a family that includes more than 20 living species. True crocodiles have longer, more V-shaped snouts than alligators and caimans do.

SIZE: Up to 6m long

DIET: Freshwater and marine fish, also reptiles, birds and mammals

FOUND IN: Marine environments as well as mangrove swamps, lagoons and estuaries in the Indo-Pacific region

SNAKES

Snakes are among the most specialised hunters on Earth. Their legless body plan means that they can squeeze into the tightest spaces, effortlessly moving through burrows and cracks to find prey. Snake senses are almost unrivalled in nature, with some species even able to sense infrared wavelengths. Of 3,600 species worldwide, not a single snake eats anything other than meat.

GREEN ANACONDA
(Eunectes murinus)

The green anaconda is the longest snake native to the Americas, sometimes reaching five metres in length. Though rather sluggish on land, the snake is a masterful hunter of wetlands and lagoons where it use its stealth to catch and feed upon a range of prey including tapirs, deer and even jaguars and caimans.

The anaconda is a representative of the boa family of snakes. Like all boas, the anaconda is non-venomous. It uses its coils to constrict prey, suffocating and killing them before swallowing them whole.

SIZE: Up to 5m

DIET: Large and medium-sized mammals, reptiles and fish

FOUND IN: Swamps, wetlands, marshes and slow-moving streams

TIMBER RATTLESNAKE
(Crotalus horridus)

The timber rattlesnake, like other so-called pit vipers, has two special holes behind the eyes, called loreal pits, which allow it to detect and home in on the body heat of its warm-blooded prey.

This species is one of the most widespread rattlesnakes in North America. Its venom contains a complex cocktail of chemicals which immobilises prey. Though this venom can be used against humans, the rattlesnake only bites as a last resort, preferring instead to use its trademark rattling tail to warn onlookers to back away.

SIZE: Up to 150cm long

DIET: Small mammals, snakes, frogs and birds

FOUND IN: Deciduous forests and rocky outcrops in the eastern USA

OLIVE SEA SNAKE
(Aipysurus laevis)

As with nearly all sea snake species, the olive sea snake is highly venomous. Its venom contains enzymes which break down fishy prey from the inside, helping the snake to digest its meal more easily. When not chasing fish, this sea snake hides amongst cracks in coral reefs. Its tail has special light receptors so that it can work out if its tail is visible to predatory sharks.

This is one of 69 species of sea snake worldwide. All of them have paddle-like tails and most rarely leave the water.

SIZE: Up to 2m long

FOUND IN: Coral reefs in Indo-Pacific seas

DIET: Fish, fish eggs, crustaceans

KING COBRA
(Ophiophagus hannah)

The king cobra is a venomous snake which can rear upwards and widen its face, producing a threatening hood that scares off predators. It is a large snake species and the longest venomous snake in the world, occasionally exceeding 4 metres in length.

Rather than hunting mammals, it uses its venom to immobilise other snakes, which it swallows whole. Though they can bite in self-defence, most king cobras are naturally quite gentle and retiring creatures.

SIZE: Up to 5.5m long

FOUND IN: Dense forests across India and Southeast Asia

DIET: Mostly snakes, but also small lizards and mammals

HIGHLY VENOMOUS SNAKES

Many snakes have hollow fangs which are capable of injecting venom into prey to stop it from running away. In many cases this venom can be incredibly powerful, often affecting the brain and heart of the victim within seconds.

INLAND TAIPAN
(Oxyuranus microlepidotus)

Drop for drop, the venom of the inland taipan is the most toxic substance produced by any snake. One bite is enough to kill 100 humans. Thankfully, this is a shy and solitary creature that normally avoids people. Instead, it uses its venom to immobilise the fast-moving mammals on which it feeds.

The inland taipan is so well adapted to its dry grassland habitat that it can change colour with the seasons. In the winter it becomes darker so that it can absorb more heat from the sun.

SIZE: Up to 2.5m long

DIET: Small mammals – mostly rats and mice

FOUND IN: The dry plains of South Australia and Queensland, Australia

BLACK MAMBA
(Dendroaspis polylepis)

The black mamba is the fastest snake on the planet, occasionally clocking up speeds of 11 kilometres per hour. Unlike other snakes, it uses its speed to chase fast-moving prey, which it takes down with a rapid strike of its venomous jaws.

The black mamba's venom is highly toxic, but it rarely uses it in self-defence. Instead, when predators approach it gives off a warning display to larger animals, opening its mouth to show off its teeth.

SIZE: Up to 3m long

DIET: Small mammals including hyraxes and bushbabies

FOUND IN: Throughout sub-Saharan Africa

STRANGE SNAKES

Not all snakes really look like snakes. Some snakes, especially so-called blind snakes, have taken to the soil and have become expert tunnel-diggers. Other snakes have developed strange flaps and trunks that make them look like something else, such as a leafy branch.

MALAGASY LEAF-NOSED SNAKE
(Langaha madagascariensis)

The Malagasy leaf-nosed snake is a true master of camouflage. When it holds its body still among the branches, it becomes almost totally invisible to predators. Its colours perfectly match those of its rainforest surroundings. It even sleeps dangling its body from branches so that it looks like a vine.

The strange protrusion that comes from this snake's nose is thought to provide extra camouflage by making its head look like a leaf.

SIZE: Up to 100cm long
FOUND IN: Dry and wet forests of Madagascar
DIET: Lizards

FLOWERPOT SNAKE
(Indotyphlops braminus)

This bizarre snake is so small it is often confused for an earthworm. It gets its common name because it is sometimes accidentally transferred between countries in flowerpots filled with soil. Females can produce babies without any males, a special way of reproducing called parthenogenesis.

The flowerpot snake belongs to a group of snakes called blind snakes. Its eyes have become little more than tiny dots behind its scales. It can't make out objects with these eyes, but it can tell night from day.

SIZE: Up to 17cm long
FOUND IN: Soil throughout Africa and Asia, though now known throughout the world
DIET: The larvae and eggs of ants and termites

LIZARDS THAT DISPLAY

Lizards are a large, varied group of reptiles. Among the group, many lizards are highly territorial. They set up special areas that they regularly patrol. If another lizard should set foot on their turf, they often show their anger by displaying a range of colours and postures.

TOKAY GECKO
(Gekko gecko)

The gravity-defying tokay gecko is one of the largest geckos in the world. Even though it weighs more than a can of soup, it can easily climb up walls using its sticky foot pads.

Today, many tokay geckos live in people's homes, where they hunt household pests. For this reason many house-owners love them, even though the male's breeding call – the loud 'TOKAY!' that gives the species its name – is enough to wake children up in the night.

SIZE: Up to 35cm long

DIET: Mostly insects and spiders

FOUND IN: Rainforests and urban areas throughout Asia

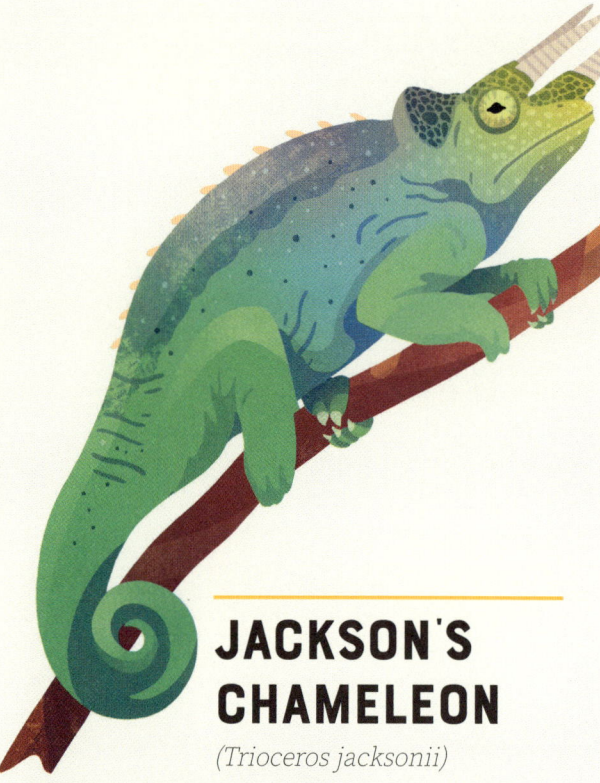

JACKSON'S CHAMELEON
(Trioceros jacksonii)

The male Jackson's chameleon has a face like *Triceratops* and a temperament to match. If he meets another male, he will pose in a special way to show off his size and change his skin colour to communicate his rage.

The Jackson's chameleon is different to most chameleons because the female gives birth to live babies rather than laying eggs. Each baby is about the size of a paper clip and could rest comfortably on the tip of a human finger.

SIZE: Up to 38cm

DIET: Mostly insects, but also centipedes, spiders and snails

FOUND IN: Native woodlands and forests of East Africa

MWANZA FLAT-HEADED ROCK AGAMA
(Agama mwanzae)

People call the Mwanza flat-headed rock agama the 'Spiderman lizard' because it shares the same distinctive colours as the superhero. Only the male has the Spiderman colouration, however, which it uses to show off to females how healthy it is.

The male is incredibly territorial and often takes part in head-bobbing contests with other males. If neither of the two lizards backs down they bare their teeth and lash out at one another with their tails. Many individuals carry wounds and broken bones from these fights.

SIZE: Up to 30cm long

DIET: Mostly insects, though they can eat berries and seeds

FOUND IN: Semi-deserts of East Africa

BROWN ANOLE
(Anolis sagrei)

The brown anole is an amazing invader. Once a popular pet, it now roams free across many parts of the United States, multiplying in numbers as it spreads.

The brown anole has excellent eyes and it is always watchful. Not only can it spot prey from many metres away, but it can also communicate with other anoles through little colourful flags that it can stick out from its neck. Like many lizards, it can shed its tail to escape predators.

SIZE: Up to 20cm long

DIET: Crickets, moths, ants and grasshoppers

FOUND IN: Originally the forests of Cuba and the Bahamas, now throughout the Caribbean and southeastern US

DEFENSIVE LIZARDS

There are more than 7,000 lizard species on Earth and nearly each one is regularly hunted by something else. For this reason, lizards have a range of ways to protect themselves and nearly all species are very watchful for predators.

COMMON FLYING DRAGON
(Draco volans)

By opening up special flaps on its ribs, the common flying dragon can glide for more than 10 metres from tree to tree. To get added control, it pulls these flaps extra-wide using its claws, a bit like a superhero in a cape. This escape mechanism has proved very successful.

Males are highly territorial. They defend particular trees and show off to nearby females with a pair of eye-catching yellow neck flaps.

SIZE: Up to 22cm long

DIET: Ants and other insects

FOUND IN: Forests throughout South India and Southeast Asia

THORNY DEVIL
(Moloch horridus)

The thorny devil is the lizard with two heads, one real and one false. When cornered by a predator, it lowers its true head and exposes behind it a 'dummy' head which is made of soft squidgy tissue. If a predator bites this, rather than its real head, the thorny devil can live to fight another day.

The thorny devil has unusual scales that can collect water droplets from the floor and slowly draw them upwards (through a process called capillary action) towards its mouth.

SIZE: Up to 20cm long

DIET: Ants

FOUND IN: Dry grasslands and deserts across central Australia

SHINGLEBACK LIZARD
(Tiliqua rugosa)

The shingleback lizard is remarkable for its mating behaviour. Males and females often stay together throughout the year and form a protective bond with one another that can last 20 years. This is very unusual for reptiles.

This lizard is well suited for life in the Australian outback. It has a chunky tail within which it stores fat that it lives off during hard times. When threatened it opens its mouth wide and shows off its bright blue tongue to scare away predators.

SIZE: Up to 40cm long from head to tail

DIET: Flowers, carrion, snails and insects

FOUND IN: Desert scrublands of southern and western Australia

GREEN IGUANA
(Iguana iguana)

The green iguana has two very sensitive eyes and a third eye (called the parietal eye) that looks upwards from out of its skull. This very simple eye can detect changes between light and dark, making it especially useful for spotting the shadows of large predatory birds that may attack from above.

The green iguana spends much of its time munching on vegetation. Its leaf-shaped teeth are very sharp and can be used for self defence. It can also strike out at predators with a powerful tail.

SIZE: Up to 1.7m long from head to tail

DIET: More than 100 different species of plant

FOUND IN: Forests and wetlands throughout South and Central America and the Caribbean

KILLER LIZARDS

Though many lizards are impressive hunters of invertebrates, some go one step further and become predators of birds and mammals. These larger lizards rely on strong jaws and some use toxic chemicals (poison) to help keep prey from running away, a bit like their reptile cousins, the snakes.

GILA MONSTER
(Heloderma suspectum)

Like snakes, the gila monster has poison glands in its jaw. It cannot inject this poison like a venomous snake would, so instead the gila monster has to chew its victims and allow the poison to flow into the bite wounds. Though this poison is highly toxic, it doesn't produce enough of it to kill most victims.

The gila monster spends much of its time underground and can survive on just five to ten meals a year.

SIZE: Up to 50cm

DIET: Mostly bird and reptile eggs

FOUND IN: Scrublands and deserts across the southwestern United States

KOMODO DRAGON
(Varanus komodoensis)

The Komodo dragon is a giant among lizards. Measuring more than twice the length of most lizards, it dominates the ecosystems of the Indonesian islands that it inhabits.

The Komodo dragon has a powerful bite which it uses to bring down and pull apart deer. Its diet also includes carrion. Using a long tongue which it flicks in and out of its mouth, it can smell a dead animal from almost 10 kilometres away. Attacks on humans have been known.

SIZE: Up to 3m long

DIET: Wild pigs and deer

FOUND IN: Forests and grasslands in some islands of Indonesia

WEIRD LIZARDS

Some reptiles look like lizards, but they are not. They include the tuatara, a reptile that has more in common with dinosaurs than modern-day lizards. Then there are the other reptiles that look nothing like lizards, but are lizards nonetheless. These include the other-worldly amphisbaenians, nearly all of which have lost their legs over time to become wormlike soil diggers.

TUATARA
(Sphenodon punctatus)

The tuatara is a lone survivor of a group of reptiles that flourished alongside the dinosaurs 200 million years ago. Unlike lizards, the tuatara has a double row of teeth on its upper jaw that slot into a single row of teeth on its lower jaw. It is the only animal on Earth to have teeth like this.

Sadly, the tuatara was almost made extinct because of rats introduced by humans to its island home. This living fossil is now on the brink of extinction.

SIZE: Up to 60cm long from head to tail

FOUND IN: Remote coastal outcrops of New Zealand

DIET: Beetles, crickets and spiders

MEXICAN MOLE LIZARD
(Bipes biporus)

The Mexican mole lizard is one of the only digging animals on Earth that is equally good at burrowing forwards and backwards. It moves in a concertina-like fashion, shifting the mud as it goes.

Like other burrowing animals, the Mexican mole lizard has dramatically reduced eyes that can only tell the difference between light and dark. Unlike many amphisbaenian species, it still has a pair of tiny forearms, which it uses to push through the soil.

SIZE: Up to 30cm long

FOUND IN: The Baja California peninsula of Mexico

DIET: Ants, termites, earthworms and other burying invertebrates

BIRDS

WHAT IS A BIRD?

Today, our planet is home to around 10,000 bird species. Birds live and feed in habitats as rich and diverse as rainforests, grasslands, wetlands, snowy mountains, deserts and even, in the case of the emperor penguin, the deep ocean. There are birds that feed on nectar, birds that feed on seeds and, as you might expect, birds that feed on other animals, especially the celebrated birds of prey.

Birds are the only surviving branch of the dinosaur family tree. Like other dinosaurs, they lay eggs with a hard shell and, like their *Tyrannosaurus* cousins, they walk on hind legs that sometimes display impressive claws. Fossils suggest that the earliest birds lived in the Jurassic Period – and it is from these early ancestors that the varied families of modern birds evolved.

Through a host of adaptations, these vertebrates have evolved warm-bloodedness, similar to mammals. This means that birds are highly active, vocal and colourful creatures that can maintain their own body temperatures. In some birds, display calls have become very complex, giving rise to what we call birdsong. In others, such as crows and ravens, warm-bloodedness has allowed the development of an impressive brain that provides a unique understanding of the world – comparable to that of apes and humans.

In many birds, males and females of the same species look very different. Nowhere is this more noticeable than in the birds of paradise, where males seek the attention of females using a wealth of rainbow colours and vivid eyespots on their feathers. Many species perform intricate dances and some even make little stages on which they display.

Throughout the world, birds have become an important part of human culture. For thousands of years, their feathers have delighted us, their songs have inspired us and their mastery of the air has enthralled and captivated us as we watch from the ground below.

BIRD CHARACTERISTICS

FEATHERS

All birds have feathers which they use to keep warm, to keep dry and to fly. There are two main types of feather: the rigid feathers required for flight and, hidden beneath them, soft downy feathers which help insulate the body. In many species, the rigid feathers are coloured, meaning that birds come in a wide range of patterns and shades. However, though some birds are bright and colourful, most birds use feather colours to camouflage from predators or prey.

WINGS

Inside bird wings are familiar dinosaur limb bones that have become adapted for flight. The bones in the lower forearms (the ulna and radius) are long and end in a set of dinosaur finger bones, most notably the so-called 'alula' which is an adapted dinosaur thumb. When slowing down or coming in to land, birds flex the alula on each wing to stretch their wings extra-wide.

EGGS

Like reptiles, birds are egg-layers. They lay shelled eggs which are nearly always incubated and cared for by parents who keep them safe and warm in the early days and weeks of life. Because of the threat of egg-predators, many birds must pair up and work as a team to look after chicks. In the case of some birds, males and females stick together in loyal pairs for years on end.

BEAKS

Birds don't have true teeth but instead use a horny beak for eating, preening and probing for food. The beak is covered in a layer of armour, called keratin. Each bird has a beak suited to its own way of life. Fish-eating birds, for instance, often have a long thin beak for stabbing at fish, while carrion-eating birds have a hooked beak for tearing prey apart.

FLIGHTLESS BIRDS

Flightless birds walk on broad muscular legs and most species use their strong beaks to pry open fruits and seeds. Though useless for flight, their small, weak wings can be used for scaring away predators or for attracting one another's attention during the breeding season.

EMU
(Dromaius novaehollandiae)

The emu can survive in some of the hottest and driest environments on Earth. To stop itself from overheating it has unique feathers that keep the body cool. Unlike many birds, the emu can go days without water. If one should stumble across a pond it can drink solidly for 10 minutes or more to stock up.

Few predators can take down an adult emu. This is partly because its large eyes and keen ears quickly sense prey coming, so it escapes before they get a chance to attack.

SIZE: Up to 2m tall

DIET: Grasses, seeds and insects including caterpillars

FOUND IN: Throughout mainland Australia

OSTRICH
(Struthio camelus)

The ostrich is the fastest creature on two legs. Powerful leg muscles and elastic-like tendons mean that as each foot hits the ground, the ostrich is catapulted more than five metres forward through the air. These adaptations mean that it can maintain speeds of more than 50 kilometres per hour, and top speeds of 70 kilometres per hour have even been known.

With speed like this, the only predator on Earth that stands a chance of bringing down the ostrich is the cheetah.

SIZE: Up to 2.74m tall

DIET: Seeds, shrubs, grasses, fruit and flowers

FOUND IN: Savannahs and grasslands of Africa

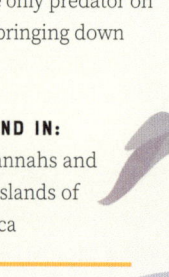

SOUTHERN CASSOWARY
(Casuarius casuarius)

When threatened by predators, the southern cassowary can lash out with sickle-shaped claws on the inner toe of each foot. These 12 centimetre long talons are among the largest of all birds' claws and resemble those of dinosaurs such as *Velociraptor*. Though potentially dangerous to humans, the southern cassowary is a secretive bird that prefers to avoid confrontation. It mostly feeds upon fallen fruits, including fruits that other animals find toxic, though it can also hunt small reptiles including snakes and lizards.

SIZE: Up to 1.8m tall

DIET: Fruits, seeds, fungus, insects and small reptiles

FOUND IN: Tropical rainforests throughout north east Australia, Papua New Guinea and Indonesia

AMERICAN RHEA
(Rhea americana)

Once a year, the male American rhea goes through a dramatic change. He suddenly becomes very territorial and begins to dig an immense hole in the ground that will become his nest. He then encourages females to lay their eggs in his nest, which he then incubates. Successful males sometimes end up with 60 or more babies.

This flightless bird has many predators. When spooked, it zigzags its way through the undergrowth using its wings as sails to keep balance.

SIZE: Up to 1.7m tall

DIET: Fruits, seeds and leaves

FOUND IN: Open grasslands throughout South America

GAME BIRDS

These are the galliformes, popularly called 'game birds'. Game birds have sharp claws that they use to kick up dust and soil whilst searching for seeds. Male game birds are very showy. Many have strange and colourful headgear which they use to display how healthy they are to passing females.

RED JUNGLEFOWL
(Gallus gallus)

The red junglefowl is the wild ancestor of the domesticated chicken. Chickens are thought to have been tamed and bred from this species more than 8,000 years ago.

In the wild, the red junglefowl is a fast-running and curious ground bird. Like many game birds, the male has a distinctive colourful flap on the head and chin and long shimmering tail feathers. Just like a domesticated cockerel, he shows off with a familiar 'cock-a-doodle-do!' cry each morning.

SIZE: 40cm tall

FOUND IN: Forests across India and Southeast Asia

DIET: Insects, seeds and fruits

WILD TURKEY
(Meleagris gallopavo)

The wild turkey is one of the world's most chatty birds. As well as purring, clucking, yelping and whining, it can produce an ultrasonic drumming sound by making air rattle around a special air sac in its chest. Its most familiar call is the gobble. A gobble can carry through the air for up to two kilometres.

Unlike many game birds, the turkey is an impressive flier. It regularly roosts in trees and can fly for more than 400 metres in open grassland to escape predators.

SIZE: Up to 125cm long from beak to tail

FOUND IN: Throughout the woodlands of North America

DIET: Seeds, berries and insects

KING QUAIL
(Synoicus chinensis)

The king quail is about as heavy as a golf ball, making it one of the world's tiniest game birds. Both males and females show a wide range of colours on their feathers, including shiny bronzes and silvers. Only males, however, have eye-catching blue feathers.

Like other game birds, the king quail has tough feet which are suited to a life of running across hard ground. Its sharp claws are used to scratch through mud and sand, exposing potential food underneath.

SIZE: Up to 14cm long

DIET: Seeds, grasses and small insects

FOUND IN: Forests through Southeast Asia and Oceania

INDIAN PEAFOWL
(Pavo cristatus)

With vibrant eyespots and shiny greens and blues, the elaborate tail of the male Indian peafowl (or peacock) is carefully inspected by the female (or peahen) who uses it to work out whether he is worthy of fathering her chicks. The peacock can even make his 200 or more tail feathers quiver, producing a strange whirring noise for added effect.

The Indian peafowl is a sociable bird. Its loud alarm call tells others in the forest of the presence of predators, including tigers.

SIZE: Up to 2.3m long

DIET: Seeds, fruits, insects and even small mammals and reptiles

FOUND IN: Forests throughout the Indian subcontinent

PIGEONS AND DOVES

Few birds have taken to our cities and towns quite like pigeons and doves. These short-billed birds excel at gathering seeds and grains, though their adaptable nature means they can make a meal of most things, including humans' leftover scraps.

ROCK DOVE
(Columba livia)

The rock dove flourishes across 10 million square kilometres of the Earth's surface, often multiplying in towns and cities in its domesticated form, the feral pigeon or town pigeon. Its success is partly down to the fact it is not at all fussy about its diet, but it is also because skyscrapers and other tall buildings are a good stand-in for the cliff habitats where it first evolved.

Depending on where it lives, the rock dove can produce eggs throughout the year. Like all doves, the rock dove provides its chicks with specially prepared milklike food that it secretes from its stomach lining.

SIZE: Up to 37cm long from beak to tail

FOUND IN: Open clifflike environments throughout the world

DIET: Seeds, fruits and human rubbish

SPOTTED DOVE
(Spilopelia chinensis)

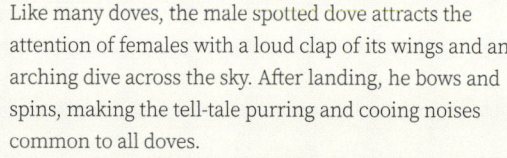

Like many doves, the male spotted dove attracts the attention of females with a loud clap of its wings and an arching dive across the sky. After landing, he bows and spins, making the tell-tale purring and cooing noises common to all doves.

Like the rock dove, the spotted dove is quickly spreading and thriving in new parts of the world. This is partly because its chicks grow very quickly. After hatching, it takes just 14 days for a baby spotted dove to fledge.

SIZE: Up to 32cm long

DIET: Grass seeds, grains and fallen fruits

FOUND IN: Common throughout the Indian subcontinent and across Southeast Asia

NEST PARASITES

Some birds look for others to take on their childcare. These birds are called nest parasites. Many nest parasites seek out particular bird species which they trick into raising their chicks unknowingly. Though risky, this sneaky behaviour can become a very clever and successful parenting tactic.

COMMON CUCKOO
(Cuculus canorus)

In springtime, the common cuckoo flies thousands of kilometres across Africa to Europe, where it begins its search for a suitable nest. It is particularly drawn to the nests of small birds including pipits and warblers.

The female cuckoo waits until the nest is unoccupied and then lays a single egg before fleeing. The cuckoo egg hatches and is cared for by the other bird parents, unaware that they are tending to a different species' chick.

SIZE: 32–34cm long from beak to tail

DIET: Insects, especially hairy caterpillars

FOUND IN: Open landscapes throughout Europe in the summer months

DUCKS, GEESE AND SWANS

Ducks, geese and swans are very vocal and can produce a range of familiar loud quacks and honks. Their webbed feet allow them to move across the surface of water, seeking out submerged weeds and leaves which they can reach down underwater to nibble.

MALLARD
(Anas platyrhynchos)

With a taste for insects, snails, pond plants and the occasional human handout, the mallard has adapted to a range of habitats, including lakes, ponds and ditches. It thrives in climates across much of the world, from Arctic tundra to tropical islands.

As with many birds, the male and female mallard look very different to one another. The female is highly camouflaged to remain hidden whilst sitting on her eggs whereas the male shows off his strength with reflective green plumage and a bright yellow beak.

SIZE: 50–65cm long

DIET: Snails and water-living invertebrates, plants

FOUND IN: Freshwaters all over the world

BLACK SWAN
(Cygnus atratus)

Like all swans, the black swan has a long and muscular neck on top of which sits a powerful beak that can be thrust deep into the water in search of food. It can also filter-feed at the water's surface, sieving out plants and algae which are then swallowed.

The black swan is incredibly protective of its family. Intruders that stray too close to its nest are quickly scared away with powerful slaps of its wings and angry bites.

SIZE: 110–142 cm long

DIET: Water plants and algae

FOUND IN: Native to Australia

BAR-HEADED GOOSE
(Anser indicus)

The bar-headed goose is a true high-flier. Scientists have tracked individuals travelling at more than 7,000 metres above sea level as they migrate between their summer and winter grounds. These epic migrations are possible because this goose has a powerful heart and giant lungs capable of taking in the thin air of the high atmosphere.

In spring and summer, the bar-headed goose congregates with others in vast colonies in the mountain lakes of central Asia. Sometimes thousands gather together in vast, noisy flocks.

SIZE: 71–76cm long

DIET: Grasses, leaves, stems, seeds and berries

FOUND IN: Southeast Asia during the winter months before migrating to Central Asia for the summer

MANDARIN DUCK
(Aix galericulata)

Springtime brings with it a magnificent transformation in the male mandarin duck. His bill reddens and special whisker-like feathers develop on his head. The feathers on his chest become purple and two sail-like patches of orange feathers develop on his wings. So elegant and magnificent is this plumage that mandarin ducks have been transported to zoos and botanical gardens throughout the world.

In China and Korea, the mandarin duck has become an emblem of good fortune, happiness and long-lasting love between couples.

SIZE: 41–49cm long

DIET: Acorns, snails, aquatic plants, aquatic insects and small fish

FOUND IN: Wetland forest habitats throughout East Asia

BRILLIANT BEAKS

Every bird's beak has a story. Some birds have curved beaks for picking through mud whilst searching for hidden invertebrates. Hooked beaks are for pulling apart nuts and seeds. Amongst the most advanced beaks are those that look like a dagger. Dagger-beaks are used to stab through water to catch unwary fish.

PAINTED STORK
(Mycteria leucocephala)

The painted stork pushes its half-open beak through the water, ready to snap it on any fish that should accidentally swim too close. This powerful beak can also be used to pin down larger wetland prey, including large frogs and snakes.

The painted stork likes to nest in the branches of trees on tiny islands where few predators can reach. Here, hundreds of storks can form noisy super-colonies. In the heat of the midday sun, adults use their long wings to keep their chicks shaded.

SIZE: Up to 102cm tall

DIET: Mostly fish, also frogs and snakes

FOUND IN: From the Himalayas to South East Asia

GREAT WHITE EGRET
(Ardea alba)

The great white egret is a large waterbird with ghostly white plumage. Unless suddenly disturbed, it rarely makes a noise. It waits like a statue at the water's edge and when prey comes too close, it shoots its dagger-like beak into the water with surgical precision.

In some parts of the world, the white feathers of the great white egret were once used to decorate hats. This fashion almost led to the bird's extinction. Thankfully, its numbers have bounced back and today it is found across many parts of the world.

SIZE: Up to 1m tall

DIET: Fish, frogs and small mammals and reptiles

FOUND IN: Wetlands throughout the Americas, Africa, Asia and Europe

GREATER FLAMINGO
(Phoenicopterus roseus)

The greater flamingo is a filter-feeder. It kicks up sand and mud on the floor of its lagoon habitat, disturbs tiny organisms and sieves them out of the water with rows of horny plates in its downward-sloping beak. Healthy flamingos filter up so many vitamin-rich crustaceans in their diet that their feathers change colour. This is what gives all flamingos their distinctive pink colouration.

Flamingos have few natural predators. Some individuals have been known to live 60 years or more.

SIZE: Up to 187cm tall

DIET: Small shrimps, blue-green algae, molluscs and seeds

FOUND IN: Wetlands throughout Africa, southern Asia, the Middle East and southern Europe

SPECTACULAR SEABIRDS

Many birds have adapted to a seafaring way of life, diving to catch fish and often spending months out at sea. Many seabirds make their nests in or on top of cliffs. Often these cliffs are on remote islands where predatory rats and foxes can't reach. These clifftop colonies can be very loud and noisy.

ATLANTIC PUFFIN
(Fratercula arctica)

In its noisy clifftop colony, the Atlantic puffin digs a burrow inside which it lays a single egg. Once hatched, the chick's mother and father will visit every few hours to deposit fish into its mouth, which it swallows greedily. Baby puffins put on weight quickly. Within six weeks a baby is grown up and ready to fledge. Once fledged, young puffins live out at sea for two or three years without once returning to land.

SIZE: 20cm tall

FOUND IN: Isolated cliffs across northern Europe

DIET: Fish, especially small eels

PERUVIAN PELICAN
(Pelecanus thagus)

The Peruvian pelican is a plunge diver. From a great height, it drops downwards upon fish, attacking them with a sturdy beak as long as a child's arm. Special air bubbles in its bones and under its skin mean that, when submerged in the water, it quickly bobs back up to the surface.

As with all pelicans, the Peruvian pelican has a throat pouch which it can spread like a net to catch many fish at once.

SIZE: Up to 152cm from beak to tail

FOUND IN: West coasts of South America

DIET: Mostly fish

WANDERING ALBATROSS
(Diomedea exulans)

In a single year, the wandering albatross may travel 120,000 kilometres as it loops round and around the Southern Ocean. This makes it one of the most wide-ranging creatures on Earth.

Each year, it travels back to its own special island where it will raise its chicks. Like many albatrosses, this species breeds with the same partner each year. Breeding pairs have been known to stay together for their whole lives, making them one of nature's most faithful birds.

SIZE: 135cm long, wingspan over 3m

FOUND IN: Islands and open sea throughout the Southern Ocean

DIET: Fish, octopus and squid

PENGUINS

With long, missile-like bodies, powerful flippers and a tapered bill, penguins deserve their status as one of the most specialised fish-hunters on the planet. Today, penguin populations are dotted throughout the southern hemisphere and between 17 and 20 species are known to scientists.

EMPEROR PENGUIN
(Aptenodytes forsteri)

The emperor penguin has been known to dive to depths of 500 metres, making it the deepest-diving bird on the planet. It has a long skeleton made of specially strengthened bones that help it sink more quickly into the murky depths as it searches for fish.

The emperor penguin has an impressive knack for endurance. In winter, each individual may make an epic 100-kilometre journey across frozen Antarctica to its nursery grounds. It is the only bird that can survive at the South Pole.

SIZE: 110–130cm tall   **FOUND IN:** Antarctica

DIET: Fish, crustaceans and squid

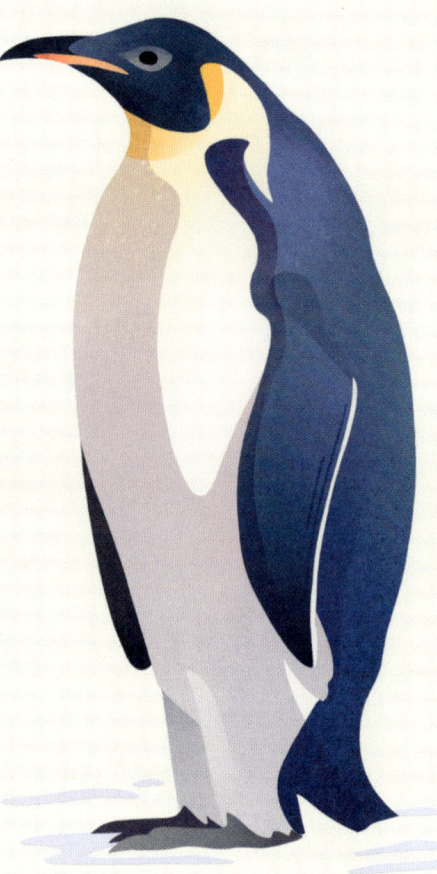

AFRICAN PENGUIN
(Spheniscus demersus)

The African penguin makes its nest out of penguin poo. Males and females sit on eggs laid in these guano (bird poo) nests for 40 days before they hatch. Later, the chicks join little crèches while the parents go off to hunt for fish.

At one time there were four million African penguins in the wild, but threats such as oil spills and illegal harvesting of fish by humans mean that just 55,000 survive today.

SIZE: 60–70cm tall   **FOUND IN:** Southwestern coast of Africa

DIET: Fish and squid

GENTOO PENGUIN
(Pygoscelis papua)

The gentoo penguin is the fastest penguin in the sea. Regularly reaching speeds of 35 kilometres per hour through the water, it pursues shoals of crustaceans (mainly krill) snapping up prey that strays within reach of its powerful beak.

Like all penguins, the gentoo penguin generates extra power with its flippers by pushing through water with both upward and downward strokes. This is different to flying birds, which generate lift with downward strokes only. Penguins' missile-like shape gives them extra streamlining.

SIZE: 51–90cm tall

DIET: Mainly krill, though sometimes fish

FOUND IN: Stony beaches and cliffs on islands in Antarctic waters

LITTLE PENGUIN
(Eudyptula minor)

The little penguin stands about as tall as a cat. Each morning it climbs out of its own little hole on the beach and enters the water for a day's hunting. At dusk, it returns with a bellyful of fish. In its nest, the little penguin will spend much of its time cleaning its feathers. On each feather it smears a tiny drop of oil taken from a special gland hidden near the tail. This oil helps keep it waterproof and streamlined while swimming.

SIZE: 33cm tall

DIET: Fish, squid and crustaceans

FOUND IN: Coasts of southern Australia and New Zealand

TORPOR BIRDS

To save energy during cold weather, a small number of birds are able to enter a sleepy state a bit like hibernation. In birds, this state is described by scientists as torpor. During torpor, the bird's breathing slows and its body temperature dips. Very little can rouse them from this deep sleep.

EUROPEAN NIGHTJAR
(Caprimulgus europaeus)

So mysterious is the European nightjar that, many years ago, people thought that it stole milk from goats. In fact, it is a nocturnal predator that hunts large insects by searching for their telltale silhouettes against the night sky.

The nightjar is able to go into a torpor during cold weather, sometimes lasting weeks on end. In this state, it rests upon the ground, becoming almost invisible to predators due to its camouflaged feathers.

Its strange call is a haunting churring noise that can be heard up to 600 metres away.

SIZE: 24–28cm long

DIET: Night-flying insects, especially moths

FOUND IN: Open country across Europe and Asia during summer, Africa during winter

BEE HUMMINGBIRD

(Mellisuga helenae)

Weighing the same as a penny, the bee hummingbird is the world's smallest bird. At just five centimetres long, it is so small it can be scared away from flowers by bees and each of its eggs is only the size of a garden pea.

Like all hummingbirds, the bee hummingbird is a powerful flier capable of performing darting and hovering manoeuvres between flowers. It does this by beating its tiny wings with impressive speed. In a single second it can flap its wings 80 times. To survive such high levels of activity, it must consume at least half its body weight in nectar each day.

To save energy when not active, hummingbirds allow their body to go into an especially deep sleep. In this state of torpor, hummingbirds can drop their heart-rate from 1,000 beats per minute to just 100 beats per minute.

SIZE: Up to 6cm long

DIET: Nectar

FOUND IN: Rainforests of Cuba

HOLE-NESTING BIRDS

Many birds seek out holes in which to nest. Some birds, such as woodpeckers and toucans, look for holes in the trunks of old trees. Others, such as kingfishers, seek out nest holes on the banks of streams and rivers.

COMMON KINGFISHER
(Alcedo atthis)

The common kingfisher patiently scans the water as it sits on its perch. If a fish shows itself, the kingfisher lunges into the water with arrow-like precision. Under the water, it pulls across its eyes a pair of see-through eyelids which act like tiny swimming goggles.

The common kingfisher nests in holes in riverbanks. It digs a burrow that points slightly upwards so that it doesn't fill up with water when flooded. Some burrows can be 90 centimetres long.

SIZE: 16cm long

DIET: Fish and some large aquatic insects

FOUND IN: Wetlands throughout Europe, parts of Asia and North Africa

TOCO TOUCAN
(Ramphastos toco)

For its size, the toco toucan's mighty bill is one of the largest of any bird. Though it looks heavy, it is mostly hollow. Inside is a long and muscular tongue used for feeding.

The exact purpose of the toco toucan's giant bill is still being investigated by scientists. It is clearly important for carving out nest holes in trees, but scientists now think it may also be used like a giant radiator to pump heat away from the body when the bird is trying to keep cool.

SIZE: 55–65cm long

DIET: Fruits

FOUND IN: Central and eastern South America

PILEATED WOODPECKER
(Dryocopus pileatus)

The nesting holes of the pileated woodpecker are so well made that, when its chicks have fledged, other animals such as raccoons and owls use them for their own nests. This house-building service means that this woodpecker plays an important job in forest ecosystems.

The pileated woodpecker has a particular taste for ant colonies. It uses its pointed beak to dig out rectangle-shaped holes in trees and then uses its long tongue to wrap around and pull out ants and their grubs.

Like many woodpeckers, the pileated woodpecker can repeatedly strike its dagger-like beak against hollow trees to make a 'rat-a-tat' drumming call. To avoid damaging its head whilst pecking so vigorously, it has special spongy parts in its skull which act as an important safety cushion for the brain.

SIZE: 40–49cm long

DIET: Burying insects, especially ants

FOUND IN: Deciduous forests throughout many parts of the USA and Canada

OWLS

Owls are among the most silent of all bird predators. Special feathers on their wings muffle the sound of their wingbeats, meaning that prey rarely hears their deathly approach.

BLAKISTON'S FISH OWL
(Ketupa blakistoni)

The Blakiston's fish owl is the world's largest owl. This giant predator sits like a statue next to water, waiting for large fish to approach and then scooping them up suddenly with dinosaur-like talons. Hunting in this way takes patience. A hungry owl may have to wait four hours or more for a single hunting opportunity.

Like many owls, the Blakiston's fish owl calls in unison with its mating partner. Together they make a long musical vibrating 'woo-woo' call that travels throughout the forest.

SIZE: 60–72cm long

DIET: Mostly fish, also amphibians, reptiles and mammals

FOUND IN: Wooded areas near water in parts of Japan, Russia and China

BARN OWL
(Tyto alba)

The barn owl is a masterful hunter of small mammals. A bowl-shaped face funnels the sounds of scurrying prey towards its sensitive ears. When it senses movement in the grass, it plunges down with claws stretched wide. Its silent wings mean that prey rarely knows what's coming.

This ghostly predator is also known for its haunting night-time screams. Each year, male and female barn owls use this shrill call to show off to one another and to mark their territories.

SIZE: 33–39cm long

DIET: Mostly small mammals, especially rodents

FOUND IN: Almost everywhere on Earth except polar regions and deserts

BURROWING OWL
(Athene cunicularia)

Unlike most owls, the burrowing owl is active both day and night. It waits on a suitable perch, scanning the open grasslands for the large insects and small mammals on which it feeds. Instead of flying after its prey, it is often spotted chasing it across the ground on busy little legs.

The burrowing owl nests in the empty burrows of ground squirrels. If disturbed in its nest, it can make a noise like an angry rattlesnake to scare away would-be predators.

SIZE: 19–28cm long

FOUND IN: Open grasslands of North and South America

DIET: Mostly small mammals and large invertebrates (including worms)

SNOWY OWL
(Bubo scandiacus)

To stay alive in the freezing tundra, the snowy owl cannot afford to be fussy. Its diet includes anything it can get its talons on, including hares, moles, mice and birds such as ducks and geese. In a single year, one snowy owl can kill as many as 1,600 items of prey.

To keep warm, the snowy owl keeps its face toward the sun on cloudless days. This helps it generate the heat required to digest prey and to gather up the strength required for the next hunt.

SIZE: 52–71cm long

FOUND IN: Arctic regions throughout North America and Eurasia

DIET: Small mammals, birds, carrion

SONGBIRDS

More than half of the world's 10,000 bird species are perching birds belonging to a group called Passeriformes, or passerines. Many of these birds are known for their loud and complicated songs, which males utter to defend territories and attract females.

CACTUS WREN
(Campylorhynchus brunneicapillus)

The loud and feisty cactus wren hops across warm grasslands and deserts, turning over leaves and twigs in search of insects and their larvae. It is so well adapted to hot climates that it never needs to drink. Instead, it gets all of its water from its food.

Unusually for songbirds, male and female cactus wrens stick together throughout the year, forming little family territories that they proudly defend with a screeching 'CHAR!' call that gets louder with each second that passes.

SIZE: 18–22cm long

FOUND IN: Southwestern USA and central Mexico

DIET: Insects, including ants, beetles and grasshoppers

COMMON STARLING
(Sturnus vulgaris)

The song of the common starling is melodic and flutelike, dotted with sections of robot-like scratching noises. Like other songbirds, it can sing constantly for a minute or more because air flows through its lungs rather than flowing in and out as with our own.

In winter, starlings gather together in huge flocks to keep safe from predators. Sometimes more than a million birds can be seen moving in a great blurry ball through the sky. These flocks (called murmurations) help starlings keep safe from predatory birds.

SIZE: 20cm long from beak to tail

FOUND IN: Native to Europe and western Asia

DIET: Mostly insects, spiders and worms

NIGHTINGALE
(Luscinia megarhynchos)

As the name suggests, the nightingale sings at night – and it is one of the world's only birds to do this. If a male cannot find a mate, he will sing a loud and expressive song filled with rich melodies which travels for hundreds of metres through the quiet night-time sky. So rich is the nightingale's song that it is has inspired the works of many great writers, including William Shakespeare and the famous Victorian poets John Keats and William Wordsworth.

As with most songbirds, the female nightingale remains mostly silent.

SIZE: 15–16.5cm long

DIET: Beetles, ants, worms and spiders

FOUND IN: Spends winter in Africa and summer in Europe and southwestern Asia

ATLANTIC CANARY
(Serinus canaria)

The Atlantic canary is an incredibly chatty bird with a soft, twittering call. Each pair defends a small patch within a large and noisy colony, and their calls stop others from stepping on their turf.

In the past, canaries were used in coal mines to help protect miners. If they suddenly went silent, this indicated to miners that poisonous gases were present, encouraging them to leave. Today the domesticated form of the Atlantic canary is a popular pet.

SIZE: 10–12cm long

DIET: Seeds of weeds, grasses and figs

FOUND IN: The Canary Islands, the Azores and Madeira

BIRD BRAINS

These large-brained birds are impressive problem-solvers that can think creatively to obtain food. So brainy are these birds and so rich are their social networks that scientists regularly refer to them as 'feathered apes'. Some species are even capable of making and using simple tools.

EURASIAN MAGPIE
(Pica pica)

For its size, the Eurasian magpie has one of the largest brains of any animal. It has been observed counting out food for its chicks and using tools to clean with. Unlike almost any other bird, an individual magpie can even recognise its own reflection in a mirror.

Even with these skills, however, life is hard for the magpie. Only one in five chicks may survive their first year. Inevitably it is the wiliest and most plucky individuals that survive to breeding age.

SIZE: 44–46cm long

FOUND IN: Northern parts of Europe and Asia

DIET: Seeds and many small animals, including birds and their eggs

NEW CALEDONIAN CROW
(Corvus moneduloides)

The New Caledonian crow is a true tool-user. With a sharp stick which it stabs into nooks and crannies, it fishes out grubs and other burrowing insects. It can also make hooks and other tools.

Because of its supersmart behaviour, this crow has been studied by many scientists eager to learn more about animal intelligence. Its secret is not only that it has a knack for trying new things, but also that these crows regularly learn from one another by carefully observing what works and what doesn't.

SIZE: 40cm long

FOUND IN: New Caledonia, an island in the South Pacific

DIET: A wide range of food, including small vertebrates, insects, snails, spiders and seeds

CALIFORNIA SCRUB JAY
(Aphelocoma californica)

The California scrub jay has an impressive memory. Every year it may hide up to 200 food items throughout its forest habitat, to come back to in winter when times become hard. But scrub jays regularly steal from one another, so this jay keeps a lookout for rivals while hiding food. If it knows it is being watched when burying food, it remembers to come back later to secretly retrieve and move its private winter snack.

SIZE: 27–31cm long

DIET: Frogs, lizards, bird chicks and eggs, insects and seeds

FOUND IN: Scrub, woodlands and gardens across the western USA and Canada

COMMON RAVEN
(Corvus corax)

The common raven is one of the only birds in the world known to play. When young, individual ravens have been observed sliding down snowbanks for fun and play-fighting with otters and wolves. As an adult, the raven swoops and circles in high winds, almost as if surfing in the sky.

The common raven is highly social. Pairs remain close to one another their whole lives, sometimes for more than 20 years. It is among the most long-lived of all intelligent birds.

SIZE: 54–67cm long

DIET: From acorns, worms and maggots to small reptiles and mammals

FOUND IN: Many habitats throughout the northern hemisphere

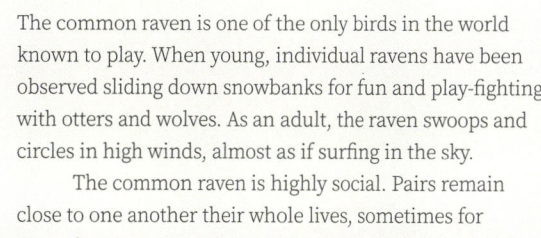

BIRDS OF BEAUTY

As well as calling loudly, many birds show off to one another using dazzling colours and patterns. In most bird species, it is males that display like this. They use their eye-catching colours as a way to communicate their health and strength to females getting ready to lay eggs.

MAGNIFICENT RIFLEBIRD
(Ptiloris magnificus)

During the breeding season, the male magnificent riflebird becomes a dazzling jewel of the forest. His chest becomes covered in shiny blue and green feathers that light up in the sun. When a female approaches he holds open his butterfly-like wings and throws his head back to hold her attention long enough for them to mate.

For many years, scientists have investigated why birds of paradise perform such intense rituals. It is likely that the male uses these dances to show off his health and strength to potential mates.

SIZE: 34cm long

DIET: Fruits, insects, spiders and other invertebrates

FOUND IN: Lowland rainforest of western New Guinea and the Cape York Peninsula, Australia

WILSON'S BIRD OF PARADISE
(Cicinnurus respublica)

The male Wilson's bird of paradise goes to great lengths to impress a mate. First, he creates a clearing by removing leaves and twigs off the forest floor. Then he sits on a branch in the centre of his stage and begins a hopping dance, spreading his reflective feathers to show off his rainbow colours. Long, jangling, metal-like tail feathers bounce up and down as he moves and his bald head shines an electric blue colour.

A female arrives. She watches the display from the branches above. If she approves, she will allow him to father her chicks before he repeats the process for another visiting female.

The Wilson's bird of paradise is among the most beautiful of the bird of paradise family, but it is an incredibly secretive bird that very few people have seen. Scientists searched for many years to capture footage of its amazing mating display on camera.

SIZE: 16cm long

DIET: Fruits, insects, spiders and other invertebrates

FOUND IN: Lowland forests on islands near West Papua, Indonesia

PARROTS

With their flashy feathers, hooked beak and their upright stance, parrots are hard to miss. In all, 400 species thrive throughout the forests and grasslands of the world, particularly in South America and Australasia.

SULPHUR-CRESTED COCKATOO
(Cacatua galerita)

To keep safe, the sulphur-crested cockatoo moves through fields and grasslands in noisy flocks. Whilst the flock is on the ground feeding, one lone bird watches out for predators from a nearby treetop. A single alarm call from this individual is enough to send the whole flock fleeing skywards.

Like many cockatoos, the sulphur-crested cockatoo is intelligent and very curious. Aside from humans, it is the only animal known to be able to shake its head in rhythm to music.

SIZE: 44–55cm long

DIET: Seeds, roots, berries and nuts

FOUND IN: Woodlands and forests of Australia and New Guinea

RAINBOW LORIKEET
(Trichoglossus moluccanus)

As with many parrots, both the male and female rainbow lorikeets are vibrant and colourful birds. They form pairs that stick together for many months. While the female feeds, the male will puff up his feathers to try and scare away love rivals.

Flowers are important to the rainbow lorikeet. Its tongue has a finger-like tip used for scooping up nectar. Like bees, it unknowingly carries pollen from flower to flower, meaning it plays an important role in forest ecosystems.

SIZE: 25–30cm long

DIET: Nectar and fruit

FOUND IN: Woodlands and forests of eastern Australia

HYACINTH MACAW
(Anodorhynchus hyacinthinus)

With its incredibly tough beak and muscular tongue, the hyacinth macaw can break into almost any nut. The tip of its beak is so sharp that it has even been known to break into coconuts. Despite this fearsome-looking beak, it is a gentle giant that rarely approaches people.

During the breeding season, it searches for holes in manduvi trees where it can make a nest strong enough to stop invading toucans from eating its chicks. Though having toucans nearby sounds alarming, the hyacinth macaw cannot live without them because toucans spread the manduvi tree seeds in their droppings. In this way, the two birds have developed an important relationship.

The hyacinth macaw was once common but has recently become increasingly rare. This is because many have been illegally captured to be kept as pets.

SIZE: 100cm long

DIET: Seeds including Brazil nuts

FOUND IN: Rainforests of central and eastern South America

BIRDS OF PREY

Many bird species play an important role as top predators in ecosystems around the world. With their streamlined wings and keen eyes, birds of prey are able to swoop down upon their unsuspecting prey animals, which they then pull apart with a hooked beak and sharp talons.

RED-HEADED VULTURE
(Sarcogyps calvus)

The red-headed vulture glides upon long wings, scanning the horizon for dead animals. Once it spots something, it flies gracefully downwards to feast upon the rotting carcass. Powerful stomach acids help it to digest decaying flesh without getting poisoned by nasty bacteria.

Like many vultures, this species has a bald head without feathers. For many years, scientists thought this adaptation was to keep its head clean but, instead, the shiny bald cap helps keep its head cool.

SIZE: 76–86cm long

DIET: Dead animals

FOUND IN: Throughout India and parts of South East Asia

ANDEAN CONDOR
(Vultur gryphus)

Pound for pound, the Andean condor is the largest flying bird in the world. Its broad wings measure more than three metres across, meaning that, once airborne, this is a bird that rarely needs to flap its wings.

Each day, it may glide 200 kilometres or more in search of dead and dying animals to feed upon. Its large size means that, when times are hard, it can sometimes bully smaller vultures away from a meal, allowing it to enjoy the spoils undisturbed.

SIZE: 100–130cm long

DIET: Mostly dead animals and occasional live rabbits and rodents

FOUND IN: Andes Mountains, South America

EASTERN BUZZARD
(Buteo japonicus)

Laser-like vision and keen ears make the eastern buzzard an impressive hunter. Rodents are an important source of food, but it is also watchful for other prey, including frogs, snakes and even worms.

As with many birds of prey, it is often approached by crows in mid-air, who peck and harass it angrily. This behaviour, called 'mobbing', lets the buzzard know it has been spotted and has lost the element of surprise, and so encourages it to hunt elsewhere.

SIZE: 40–46cm long

DIET: Mostly small mammals, birds and reptiles, occasionally large invertebrates

FOUND IN: Forests and woodland edges throughout Mongolia, Japan and China

OSPREY
(Pandion haliaetus)

The osprey thrives on every continent except Antarctica. It specialises in hunting fish, which it plunges down upon with dagger-like talons. Its toes are covered in tiny spines that help it grip its slippery meal.

This bird of prey has a number of impressive adaptations for hunting in water. Like penguins, the osprey has oily feathers to help keep buoyant and, like dolphins, it can close its nostrils when it dives to stop it from breathing in water.

SIZE: 60cm long

DIET: Fish

FOUND IN: Wetland habitats throughout the world

PHILIPPINE EAGLE
(Pithecophaga jefferyi)

Nothing terrifies monkeys quite like the sight of a Philippine eagle, perhaps the largest eagle in the world. Like a tiger, this apex predator moves silently through the rainforest, its eyes alert for the giveaway movements of tree-living mammals, especially monkeys.

To catch their prey successfully, Philippine eagles will sometimes hunt in pairs. One will sit ominously close to a troop of monkeys to catch their full attention while the other comes in from behind to deliver a surprise attack.

SIZE: 86–102cm long

DIET: Mammals, including monkeys, deer and even bats, and reptiles

FOUND IN: Rainforests of the Philippines

NORTHERN GOSHAWK
(Accipiter gentilis)

The northern goshawk is a perch-hunter. Sitting hidden high up in a tree, it scans its surroundings for hunting opportunities. Once it spots prey, it darts in for the kill.

Its short wings and long tail allow it to weave and twist acrobatically through dense woodlands as it makes its stealthy approach. After capturing its prey, it may return to a special 'plucking perch' where it pulls apart its food with its tough beak and claws before eating it.

SIZE: 46–59cm long

DIET: Mammals and medium- to large-sized birds

FOUND IN: Many undisturbed woodlands and forests throughout the northern hemisphere

PEREGRINE FALCON
(Falco peregrinus)

No other creature on Earth can defy gravity quite like the peregrine falcon, the fastest animal on the planet. This agile bird of prey regularly hits speeds of up to 389 kilometres per hour as it dives downwards on unwary birds. The resulting impact is hidden within a flurry of feathers. Death is almost instantaneous.

To cope with the wind generated as it flies down, the peregrine falcon has special bone-filled nostrils which guide air over, rather than into, its lungs. A third pair of eyelids has become a useful set of flying goggles.

In recent times, the peregrine falcon has become a familiar resident of many towns and cities. Urban buildings, especially churches and skyscrapers, resemble the clifftop habitats where they normally nest, and peregrines have become an important predator of urban pigeons and doves.

SIZE: 34–54cm long

DIET: Mostly medium-sized birds, including songbirds, pigeons and doves

FOUND IN: Almost every habitat on Earth except high mountains, rainforests and polar regions

WHAT IS A MAMMAL?

Since the end of the dinosaurs, mammals have gone from strength to strength. Today, more than 6,000 different species live on Earth. Among the smallest are the Etruscan shrew and the bumblebee bat. The largest is the blue whale: it is larger than any animal that has ever lived on our planet. Mammals thrive in forests, jungles, deserts and snowfields. Among their ranks are many of the world's apex predators, including the tiger, the lion, the polar bear and the squid-hunting giant, the sperm whale.

Mammals come from an age before dinosaurs, having evolved from reptile-like ancestors more than 300 million years ago. Some parts of the mammal group still have some reptile features. For instance, monotremes (which include the echidna and platypus) still lay eggs with a hard shell. Most mammals, however, give birth to live young which the mother provides with milk to help them grow. Some mammals keep their babies in a special pouch during the early days of their development. These are called the marsupial mammals.

One of the secrets behind the success of this group of vertebrates is their warm-bloodedness. This is a feature that helps them cool or warm their body according to their surroundings. Warm-bloodedness gives mammals a competitive advantage over other groups of animals when adapting to new habitats. This feature also allows mammals to power a bigger brain.

Nearly all mammals have well-developed brains that help them to explore and understand their surroundings through sight, sounds, smells and touch. In many mammal species, particularly monkeys, apes and humans, this big brain allows for the development of complex societies rich in culture and communication.

MAMMAL CHARACTERISTICS

HAIR
Most mammals are covered in cylinder-shaped, scalelike structures we commonly call fur or hair. This hair traps a layer of warm air next to the body just like a woolly jumper, helping mammals keep warm in cold environments. Mammal hair can be a range of colours and this often helps mammals with camouflage and communication.

WHISKERS
Most mammals have clusters of especially long hairs on their face that provide information to the brain about unseen surroundings. These are called whiskers. Nocturnal animals use whiskers to navigate their tunnels at night.

TEETH
Nearly all mammals have very distinctive jaws made up of flat teeth at the front, alongside pointy teeth with flatter teeth at the back used for chewing. In nearly all mammals milk teeth are shed to make way for larger teeth that grow into adulthood. However, in some mammals, including rodents and elephants, teeth can continue to grow throughout life.

MILK
All mammal babies are given milk by their mother as a way to help them grow quickly. This milk is highly nutritious. In some species, especially of marine mammals, milk can contain as much as 60 per cent fat. This is far more calorific and energy-rich than even the most expensive ice cream.

LIVE BIRTH
Most mammals give birth to live young. They have some sort of belly button – a scar where a tube called the umbilical cord connected them, as a developing embryo, to their mother. But there are two groups of mammals that don't. One is the marsupials, whose young are born early as tiny larvae that must crawl through fur to their mother's pouch where their growth continues. The other group is the monotremes, which are famous for being mammals that lay eggs.

OLD WORLD MONKEYS

Old World monkeys get their name because they are found in what was once called the 'Old World' – Africa, Asia and Europe. Most Old World monkeys are medium and large monkey species that live in trees, although some, like baboons and mandrills, spend much of their lives on the ground.

MANDRILL
(Mandrillus sphinx)

The mandrill moves through the rainforest in noisy groups called hordes, looking for fruits and seeds. Sometimes these hungry hordes can number more than 850 individuals. Fights are very common in mandrill groups, and so each individual will pull a special face to keep things calm. The redness of their face and bottom signals their status within the group.

The mandrill can weigh up to 37 kilograms, which is eight times heavier than an average domestic cat. It is the largest of all Old World monkeys.

SIZE: 70–100cm long including its short tail

FOUND IN: Tropical rainforests throughout Central Africa

DIET: Mostly fruits and insects, also leaves and bark

KING COLOBUS
(Colobus polykomos)

The king colobus has a stomach a bit like that of a cow. Part of its gut is enlarged to allow food to be fermented (softened) by bacteria before digestion. This allows it to feed on tough leaves that most monkeys cannot eat.

The king colobus is also important for spreading the seeds of various trees around the forest. Undigested seeds are passed out in its droppings, helping these plants colonise new places.

SIZE: 67cm long plus tail up to 90cm long

FOUND IN: Lowland and mountain rainforests in West Africa

DIET: Mainly leaves, also fruit and flowers

JAPANESE MACAQUE
(Macaca fuscata)

No monkey can withstand freezing temperatures quite like the Japanese macaque. Its fur is incredibly thick, meaning that even if the temperature drops to minus 20 degrees Celsius, its body remains warm and snug. If things get too cold, it may even look for a hot spring in which to have a warm bath.

The Japanese macaque is also one of the cleverest monkeys. As well as learning how to clean and prepare food, it regularly plays. It is the only monkey known to make and throw snowballs.

SIZE: 52–57cm tall (average)

DIET: Fruits, leaves and fallen seeds

FOUND IN: A range of forest habitats in Japan, including subtropical forests and northern Arctic-like forests

SACRED LANGUR
(Semnopithecus entellus)

The sacred langur is one of the most chatty of all monkeys. It can make a range of sounds including barks, coughs, whoops, grunts, pants, honks and hiccups.

Langur monkeys are considered sacred animals in the Hindu religion and many live alongside humans in towns and cities across India. During the day, they wander the streets in small groups looking for leaves and seeds and, occasionally, human handouts. They can make impressive leaps between buildings, sometimes jumping four metres or more in a single bound.

SIZE: 120–180cm long including tail

DIET: Fruits, grasses, flowers and leaves

FOUND IN: Rural and forested areas, as well as some towns and cities in southwest and central India

MINI MONKEYS

Monkeys and apes belong to a mammal group called the primates. But there is another often forgotten part of the primate group – the prosimians. Many prosimians are small and live a secretive life in the trees. Most have large, watchful eyes to help them spot predators including eagles, owls and snakes.

BROWN GREATER GALAGO
(Otolemur crassicaudatus)

As the sun sets, the brown greater galago wakes up ready to patrol its territory. Under the cover of darkness it moves through the trees, searching out fruits and seeds. Its highly sensitive ears twitch furiously as it listens out for the sounds of approaching predators.

This primate is incredibly protective of its feeding patch. It spends much of its time advertising its presence to rivals using a smelly scent produced by a special gland on its chest.

SIZE: 55–102cm long including tail

FOUND IN: Forests of southern and eastern Africa

DIET: Berries, figs and seeds

PHILIPPINE TARSIER
(Carlito syrichta)

The Philippine tarsier has eyes so big they are rooted into its skull. This makes looking around very difficult, so it has a special neck like that of an owl, which allows it to swivel its head around up to 180 degrees. For its size, it has the largest eyes of any mammal.

It is a keen jumper, sometimes springing three metres from tree to tree. Circle-shaped pads on its feet help it to grip onto branches tightly when it lands.

SIZE: 85–160mm tall

FOUND IN: Rainforests of the Philippines

DIET: Insects and spiders

BENGAL SLOW LORIS
(*Nycticebus bengalensis*)

The Bengal slow loris is one of the world's only venomous mammals. Its saliva contains more than 100 toxic chemicals, which it delivers to other animals through a sharp bite. Slow lorises often use this venom on one another, particularly if a rival tries to steal territory, but they can also use their bite to defend themselves from predators, especially cats.

Bengal slow lorises move through the forest in small family groups that sleep together during the day cuddled up in holes in trees. As with many primates, family members regularly groom one another to keep clean and healthy.

During the night they search for food with big eyes that can see in the dark. A special layer of tissue in each eyeball reflects light inwards so that they can scan their rainforest habitat on even the most moonless night.

SIZE: 26–38cm long from head to tail

DIET: Mostly fruits and, in winter, sugary gum that leaks from some trees

FOUND IN: Rainforests across part of the Indian subcontinent

LEMURS

Around 90 million years ago Madagascar split from mainland Africa, taking with it a clutch of primates that would evolve into its own special group – the lemurs. Today, lemurs are agile and resourceful and include perhaps the weirdest primate of all, the aye-aye.

RING-TAILED LEMUR
(Lemur catta)

The ring-tailed lemur is incredibly social. In groups 30-strong it roams the forest looking for fruits and leaves. Unlike many lemurs, the ring-tailed lemur regularly rears up and walks on its hind legs when provoked or if faced with predators.

As with most lemurs, the female ring-tailed lemur is the boss. She maintains order in the group by chasing, cuffing and biting troublemakers. When new food sources are discovered, it is she who will decide at what point the males are invited to eat.

SIZE: 95–110cm long including tail

DIET: Mostly leaves and fruits, also insects and small vertebrates

FOUND IN: Forests and scrublands of south and southwestern Madagascar

AYE-AYE
(Daubentonia madagascariensis)

The aye-aye is a nocturnal lemur like no other. Using its fingers, it taps on the sides of trees listening out for hollow bits where grubs may be hiding. When it finds such a hollow piece of wood, it uses sharpened front teeth to gnaw downwards so it can get closer. Then it takes its long middle finger, which looks like a spider's leg, inserts it into the hole to hook its meal, then pulls it out and eats it. The aye-aye gets away with this lifestyle because there are no species of woodpecker on Madagascar to compete with it.

Even though it is the largest and heaviest nocturnal primate in the world, the aye-aye manages to spend most of its time in the delicate branches of the forest canopy. Like a squirrel, it jumps from trunk to trunk, sometimes travelling four kilometres or more in a single night.

SIZE: Up to 100cm long including tail

DIET: Seeds, fruits, nectar, fungi and insect larvae

FOUND IN: Forests of eastern Madagascar

NEW WORLD MONKEYS

The 'New World' was the name given to the Americas when they were 'discovered' by European people around 500 years ago. New World monkeys, so called because they are found in North and South America, cover a range of sizes and styles. Many are incredibly adapted to a climbing lifestyle and some can even use their tail as a fifth limb.

WESTERN PYGMY MARMOSET
(Cebuella pygmaea)

The pygmy marmoset is a gummivore, which means its diet mainly consists of the sugary gums and saps released by trees. It uses specialised teeth to gnaw little holes in bark that soon fill up with sugar-rich fluid. Sometimes, it supplements its diet with butterflies attracted to the scent of this fresh sap.

Like all marmosets, the pygmy marmoset has claws rather than nails. It also has strange whiskers on its wrists through which it can feel branches and twigs.

SIZE: 30–35cm long including tail

FOUND IN: Rainforests of the western Amazon

DIET: Gums and saps

BROWN HOWLER MONKEY
(Alouatta guariba)

The brown howler monkey is one of the loudest animals on Earth. Males and females group together in the treetops and make whooping calls that can travel a kilometre or more. These territorial calls warn other howler monkeys to move on.

Like all howler monkeys, this species has a muscular prehensile tail that provides it with a fifth leg to grasp onto trees and branches. A bald spot on the end of the tail provides extra grip. Each individual brown howler monkey has its own unique fingerprint on this bald patch.

SIZE: 93–126cm long including tail

FOUND IN: Rainforests of South America

DIET: Fruits and leaves

CRESTED CAPUCHIN
(Sapajus robustus)

The shells of fruits, nuts and seeds are no match for the crested capuchin and its impressive fingers. With its dexterous hands it can carefully tease apart these food sources to get to the nutritious parts within.

Like all capuchins, this one is very sociable. Groups of up to 20 individuals roam the treetops looking for feeding opportunities. If spooked by a predator, a crested capuchin releases a special call (called the 'scream embrace mechanism') which makes all members of a group quickly come together to seek safety.

SIZE: 104-173cm long including tail

DIET: Fruits, seeds, insects and occasionally small vertebrates

FOUND IN: Lowland tropical forests and some isolated dry forests of eastern Brazil

BLACK-HEADED SPIDER MONKEY
(Ateles fusciceps)

The black-headed spider monkey has incredibly long arms and hooklike fingers which it uses to swing between branches as it moves through the forest in search of food. Like all spider monkeys, it has an impressive memory and can remember exactly when and where in the forest the ripest fruits are likely to be blossoming.

Because of hunting and the destruction of its habitats, the black-headed spider monkey has become one of the rarest of all monkeys. Today it is listed by scientists as endangered.

SIZE: 110–140cm long including tail

DIET: Mostly fruits and nuts

FOUND IN: Forests throughout Colombia, Nicaragua and Panama

APES

Apes are a tailless part of the primate family that includes both the largest primate, the gorilla, and the most intelligent primates, including the chimpanzee and the bonobo. All apes are capable of impressive feats of communication and they can even make and prepare simple tools.

BONOBO
(Pan paniscus)

Along with the chimpanzee, the bonobo is the closest living relative of modern humans. It was discovered in 1928 by museum scientists who noticed that skulls thought to have been from a chimpanzee were in fact a different species.

Compared to the chimpanzee, the bonobo is a much more peaceful ape. To reduce tension within the group, individuals regularly groom, massage and even kiss one another. Like humans, the bonobo is even fond of tickling. Its laugh is a little bit like that of a human baby's.

SIZE: 70–83cm from nose to tail

DIET: Mostly fruit, also leaves, honey and even monkeys

FOUND IN: In the humid forests of the Democratic Republic of Congo

SUMATRAN ORANGUTAN
(Pongo abelii)

The Sumatran orangutan is a master tool-crafter. It collects and sharpens special sticks which are used for various purposes. Some sticks are blunt at the end for pulling out honey from bees' nests. Others are long and sharp and are used for teasing termites out of tree holes. Orangutans can also use leaves as umbrellas.

Like all orangutans, the Sumatran orangutan clambers through the treetops. It is the heaviest mammal in the world to do this.

SIZE: Up to 180cm tall

DIET: Mostly fruits and insects

FOUND IN: Only in the northern forests of Sumatra, Indonesia

WESTERN GORILLA
(Gorilla gorilla)

Standing almost six feet tall and with an armspan of almost three metres, the silverback (mature adult male) western gorilla is a force not to be reckoned with. He is the leader and guides members of his group through the forest in search of food, as well as protecting them from leopards. But life is hard for silverbacks. Each one is regularly challenged by rival males, leading to dramatic fights.

Like other apes, the western gorilla is highly intelligent. Individuals laugh and play together, and scientists have seen them seeking particular plants which they use as medicine.

The future for this gorilla is uncertain. As well as their forest habitats being lost, they are regularly poached.

SIZE: 135–155cm tall (silverbacks can be 175cm tall)

DIET: Mostly leaves, fruit, flowers and bark

FOUND IN: Mountain forests, lowland forests and swamps in western regions of Central Africa

BATS

Bats are the only mammals to have mastered flight. They fly with wings that are lengthened mammal finger bones and their skeletons are dramatically lightened to allow for a life in the air. Most bats are nocturnal, in contrast to their rival flying vertebrates, the birds, which can fly by day.

LARGE FLYING FOX
(Pteropus vampyrus)

The large flying fox has a skull like a crash helmet. On strong wings, it crashes into the rainforest canopy, grabbing at branches with its claws before tearing fruits apart with its sharp teeth. Sometimes it can hurt itself during these crashing dives, but it can heal itself well.

This bat also drinks nectar from flowers. It has a long, hairy tongue that it keeps rolled up deep within its ribcage when not in use.

SIZE: Up to 150cm wingspan

FOUND IN: Forests across much of Indonesia

DIET: Fruits and nectar

COMMON VAMPIRE BAT
(Desmodus rotundus)

In the depths of night, the common vampire bat lands on large mammals such as cows and horses and makes a small gash in their skin with its teeth before lapping up their blood with its tongue. Like mosquitoes, the vampire bat has saliva which contains a chemical to stop its blood meal clotting up.

When back at the nest, vampire bats often share out their meals by regurgitating food for others to drink. This makes them some of the most co-operative animals on Earth.

SIZE: 18cm wingspan

DIET: The blood of large mammals

FOUND IN: Throughout parts of Mexico, Central America and South America

LESSER HORSESHOE BAT
(Rhinolophus hipposideros)

Weighing almost as little as a sheet of paper, the lesser horseshoe bat is one of the world's smallest bats. Like a speeding spaceship, it dips and dives between leaves and branches in search of small insects.

Horseshoe bats have special leaflike noses that they use to direct the sounds used in echolocation. This makes them very efficient predators. When not hunting, this species spends much of its time in busy roosts in caves, attics and barns.

SIZE: 19–25cm wingspan

DIET: Flies, moths and spiders

FOUND IN: Warm forested hilly regions throughout Europe and parts of the Middle East

SPOTTED BAT
(Euderma maculatum)

For its size, the spotted bat has the largest ears of any mammal. It uses them to listen out for grasshoppers, which it destroys with a swift bite of its jaws.

The spotted bat is part of a large group of bats called the vesper bats. While some bats direct their ultrasonic noises at prey like a spotlight, vesper bats hunt by letting their noises echo all around. Their large ears pick out the sound reflections from lots of directions at once.

SIZE: 35cm wingspan

DIET: Grasshoppers and moths

FOUND IN: Cracks and caves along Arizona's Grand Canyon, USA

WHALES AND DOLPHINS

Whales and dolphins live their whole lives in the water. They have streamlined bodies and a powerful tail to propel their bodies forwards. They have enormous brains, complex social behaviours and an impressive ability to hunt prey by detecting reflected sound waves.

ORCA
(Orcinus orca)

The orca (sometimes called the killer whale) is a spectacular apex predator. It hunts prey in well-co-ordinated attacks that often span many kilometres, so it is no wonder some people call it the wolf of the oceans.

Despite its other name, the orca is a dolphin, not a whale, and like other dolphins it uses echolocation to find food. It fires intense bursts of sound at potential prey, listening out for reflected sounds which are picked up by the brain through a lower jaw that works a bit like an ear. Like apes, the orca has a complex and very rich social life.

SIZE: 5–8m long from nose to tail

FOUND IN: Across all of the world's oceans

DIET: Marine mammals, sea turtles, fish, squid and seabirds

SPERM WHALE
(Physeter macrocephalus)

The sperm whale is perhaps the only animal big enough to hunt and kill the giant squid. Hunts take place in the deep sea and can be very violent. An adult sperm whale often bears scars on its face from struggles with this monster-like prey.

To dive down to the intense pressure of the deep ocean, the sperm whale has a ribcage that can collapse. It can hold its breath for 90 minutes or more.

SIZE: 11–20m long

DIET: Squid, octopus and occasionally deep-sea rays and sharks

FOUND IN: All of the world's oceans, preferring ice-free waters more than 1,000m deep

BLUE WHALE
(Balaenoptera musculus)

Bigger than any dinosaur or prehistoric shark, the blue whale may be the largest animal ever to have lived on Earth. Its body is 30 times as heavy as a *Tyrannosaurus rex*. Its tongue alone weighs as much as an elephant. Forty per cent of its weight comes from the powerful muscles that power its broad tail.

The blue whale reaches such a huge size by consuming clouds of tiny shrimp-like crustaceans called krill. Each day it eats up to 40 million of these creatures, which it sieves out of vast mouthfuls of water. To find enough food to survive, it must spend much of its time travelling the oceans. In a single day, an adult can cover 450 kilometres.

The blue whale was almost hunted to extinction by humans until 1966 when it became illegal for people to hunt and catch them.

SIZE: Up to 30m long

DIET: Krill and other tiny crustaceans

FOUND IN: Oceans across the world

RODENTS

About 40 per cent of all mammals alive today belong to a group called the rodents. Rodents have adapted to a host of unusual habitats including scorching deserts, barren mountaintops and freezing lakesides. Some, such as the naked mole-rat, even spend their whole lives underground and never see the light of day.

LONG-EARED JERBOA
(Euchoreutes naso)

Desert environments have very little by way of food, so the long-eared jerboa has enormous ears that for its size are among the largest of any animal. It uses them to listen for the giveaway noises of insect prey. When it detects movement it powers itself across the floor, springing off long bones in its feet like a carnivorous micro-kangaroo. Its long tail helps it swerve and dodge the attentions of its mortal enemy, the little owl.

SIZE: 22–25cm long

DIET: Insects

FOUND IN: Isolated deserts and dry habitats of Central Asia

SHORT-TAILED CHINCHILLA
(Chinchilla chinchilla)

The short-tailed chinchilla has the thickest fur of any land mammal. This helps it survive the cold mountain nights during which it searches for food. Unusually for a small mammal, it moves in herds of up to 100 individuals. By grouping up with others in this way, it can protect itself from predators including owls, skunks and snakes.

During the day, it sleeps in a burrow dug into the ground or made between crevices in rocks.

SIZE: 42.5–47.5cm long

DIET: Seeds, roots, grasses and fruits

FOUND IN: The Andes Mountains, Chile

NAKED MOLE-RAT
(Heterocephalus glaber)

The naked mole-rat is a mammal like no other. Like ants and wasps, it lives in a nest that is dominated by a single queen rat who is responsible for producing all of the babies. To survive in its underground world, the naked mole-rat has special lungs able to work in low oxygen environments.

Scientists are very interested in the naked mole-rat because it is one of the only animals in the world that does not suffer from cancer.

SIZE: 11–13.5cm long, including tail

FOUND IN: Tropical grasslands of East Africa

DIET: Plant roots

AFRICAN PYGMY MOUSE
(Mus minutoides)

The African pygmy mouse collects water in an unusual way. Every evening it stacks cold pebbles in front of the entrance to its burrow and these collect dew throughout the night. When the mouse wakes up, it drinks this dew for its breakfast.

This mouse is so tiny it could rest comfortably on an adult human thumb. It grows incredibly quickly. A baby female is ready to become pregnant herself within about 60 days of being born.

SIZE: 5–12cm including tail

DIET: Grass seeds and small insects

FOUND IN: Throughout the deserts of sub-Saharan Africa

SUPERSIZED RODENTS

Rodents have a pair of sharp incisor teeth that continue growing throughout life. Though many people think rodents are mouse-sized, and there are hundreds of small rodent species, some rodents are much bigger.

CAPYBARA
(Hydrochoerus hydrochaeris)

The capybara is the largest rodent in the world. In groups of up to 100 individuals, it meanders through the wetlands of South America searching out grasses and aquatic plants to nibble through with its especially large front teeth. When spotted by a leopard, it dives underwater and can hold its breath for five minutes.

Like many rodents, the capybara will sometimes eat it own droppings to make sure every bit of nutrition in its diet is digested. This behaviour is called coprophagy.

SIZE: 106–134cm long

DIET: Aquatic plants and grasses, fruits and tree bark

FOUND IN: Dense South American shrublands and forests near water

EURASIAN BEAVER
(Castor fiber)

The Eurasian beaver is capable of an impressive feat of engineering. By using sticks to dam up streams and rivers, it creates a pond or lake in which it builds its floating nest of sticks. This becomes its dwelling – a floating safe-house in which it can come and go without fear of being spotted by predators.

A century ago, the Eurasian beaver was almost hunted to extinction for its fur and only 1,200 were left. The species is bouncing back thanks to scientists carefully reintroducing them back into the places where they once thrived.

SIZE: 105–150cm long from head to tail

DIET: Thick-stemmed wetland plants, tree bark

FOUND IN: Isolated wetlands throughout Europe and Asia

SUPER-SUCCESSFUL RODENTS

Because they are quick to reproduce, some rodents can multiply their numbers very quickly if they should hit upon a free food source. In some cases, these rodents can accidentally be moved to different parts of the world where they have no predators and so thrive.

BROWN RAT
(Rattus norvegicus)

The brown rat originally lived in northern China but after developing a taste for human scraps, it spread. It now lives near human settlements across nearly all of the world except Antarctica. Alongside humans, this is perhaps the world's most successful mammal.

Though many people consider it a pest, the brown rat is a highly intelligent and social mammal. Rats regularly care for and groom one another and scientists recently discovered that, when tickled, the brown rat can let out a high-pitched chirp in delight.

SIZE: 15–27cm long plus tail 10.5–24cm long

FOUND IN: Near human habitations all over the world except Antarctica

DIET: Almost anything – the brown rat is an incredibly varied omnivore

EASTERN GREY SQUIRREL
(Sciurus carolinensis)

The Eastern grey squirrel is able to scurry down trees head first. It manages this with backward-pointing claws on its hind feet. When on the forest floor it spends most of its time hiding nuts and seeds which it returns to when food becomes scarce in winter.

This squirrel originally lived in North America, but some were released into Europe more than 100 years ago and have since flourished, forcing out the native red squirrel from many of the places it used to live.

SIZE: 42–55cm including tail

DIET: Seeds, nuts and tree bark

FOUND IN: Mature woodlands, parks and gardens in North America and much of Europe

HOPPING MAMMALS

Like rodents, the 'lagomorphs' (rabbits, hares and pikas) have long incisor teeth that continuously grow throughout life. With athletic back legs capable of catapulting the body forwards, most are very good at escaping from predators.

SNOWSHOE HARE
(Lepus americanus)

The snowshoe hare gets its name from a pair of extra-large feet, which spread out its weight while it hops so that it doesn't sink into soft snow. Its feet are specially adapted so they don't freeze and its soles are covered in thick fur.

To keep itself hidden from predators, the snowshoe hare can change colour throughout the year. In winter its fur is white to camouflage against the snow and in summer it becomes an earthy brown colour to hide among grasses and soil.

SIZE: 41–51cm long from nose to tail

FOUND IN: Forests throughout cooler regions of North America

DIET: Branches, twigs and small-stemmed plants

AMERICAN PIKA
(Ochotona princeps)

When food is plentiful the American pika hides food in specially selected piles of grasses. This behaviour is called 'haying'. In the peak of summer, it may hay its food up to 100 times in a single day – more than 13 times an hour!

It guards its territory closely and is very careful to steer rivals away from its food stashes. In the spring, pikas become friendlier. Males and females are even known to sing to one another.

SIZE: 16–22cm long

FOUND IN: Mountains of western North America

DIET: Plants including grasses, sedges and thistles

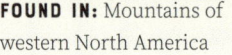

SUMATRAN STRIPED RABBIT
(Nesolagus netscheri)

The Sumatran striped rabbit is one of the most mysterious and elusive rabbits in the world. To capture photographs of this species in the wild, scientists use special cameras which can be set off by movement at night. Only a handful of real-life sightings have ever been reported.

The Sumatran striped rabbit is likely to be a very fearful nocturnal herbivore that seeks out the unused burrows of other larger digging mammals to keep safe. Scientists are working hard to find out more about this unusual mammal species.

SIZE: 42cm including tail

DIET: Plants stalks, twigs and leaves

FOUND IN: The forests of western Sumatra in Indonesia

BLACK-TAILED JACKRABBIT
(Lepus californicus)

To survive in its warm desert habitat, the black-tailed jackrabbit has an enormous set of ears lined with special blood vessels that pump heat away from the body. This helps it keep cool whilst it moves in search of food.

This rabbit has large eyes that are highly sensitive to predators, including birds of prey hovering from above. Its young are born with their eyes fully developed, ready to watch for enemies. Within minutes, these tiny babies are able to walk and run for cover if spotted.

SIZE: 61cm long from head to tail

DIET: Mostly shrubs and grasses

FOUND IN: Throughout the western USA and Mexico

SEALS AND SEAL-LIKE MAMMALS

With powerful jaws and keen eyes, seals and their close relatives are active hunters of fish. They have flippers and swim in the sea but also spend time on land. They belong to a large group of predatory meat-eating mammals called the carnivorans.

WALRUS
(Odobenus rosmarus)

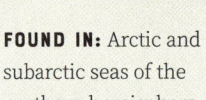

The walrus has long tusks which are actually its canine teeth. These can be used like ice picks to help it pull itself out of the water. In males, the tusks can also be used to show off to and scare away rivals. Sometimes they can reach a metre or more in length.

The walrus hunts by pushing its snout through the sandy seabed. Sensitive whiskers detect the movement of submerged molluscs which it quickly digs out and swallows.

SIZE: 2.2–3.6m long from nose to tail

FOUND IN: Arctic and subarctic seas of the northern hemisphere

DIET: Mostly submerged molluscs, also shrimps, crabs, soft corals and tube worms

CALIFORNIA SEA LION
(Zalophus californianus)

The California sea lion is very noisy. On land, its colonies are filled with cacophonous barks, grunts, belches and growls. Many of these calls are used to stop fights breaking out, particularly between males.

Like all sea lions, the California sea lion has ear flaps and long, powerful front flippers. It can also walk for short distances across sand using all of its four limbs. When hunting fish it can use a variety of clever tactics, including watching and following pods of hungry dolphins.

SIZE: 1.8–2.4m long

FOUND IN: Coastlines of western North America

DIET: Mainly fish and squid

ANTARCTIC FUR SEAL
(Arctocephalus gazella)

Using its well-developed eyes and other keen senses, the Antarctic fur seal seeks out fish, krill and squid. In a single year, one individual may eat as much as 1,000 kilograms of prey. This species is even known to eat penguins.

Unlike other seals, the Antarctic fur seal is a solitary hunter that, apart from when it rears young, rarely returns to land. In fact, a juvenile may spend years at sea without ever once coming to shore.

SIZE: Up to 2m long

DIET: Squid, krill, fish and penguins

FOUND IN: 95 per cent of all Antarctic fur seals breed on the South Atlantic island of South Georgia

SOUTHERN ELEPHANT SEAL
(Mirounga leonina)

In elephant seal society, the beachmaster is king. This giant male protects lots of females and is regularly challenged for his place by other males who provoke him to energy-sapping wrestling bouts. A long, bulbous nose gives the beachmaster an extra-loud roar.

While searching for food, this seal regularly dives down to extreme depths. Scientists have clocked this species diving to 2,133 metres. This makes it one of the world's deepest-diving mammals and is four times deeper than the deepest-diving penguin.

SIZE: Up to 5.8m long

DIET: Squid and fish

FOUND IN: Throughout subantarctic waters

ELEPHANTS AND SIRENIANS

Though they look very different, elephants and sirenians evolved from the same part of the mammal family tree more than 50 million years ago. Both of these mammals are long-lived and can move over long distances. Each has an impressive memory for where and when food sources are to be found.

ASIAN ELEPHANT
(Elephas maximus)

The Asian elephant is smaller than its African cousin. Its body slopes downwards from the head and it has smaller ears. Like all elephants, it communicates using noises so deep that humans cannot hear them – called infrasound. It can even communicate by feeling vibrations made by sound moving through the ground.

The Asian elephant is often found near water. To sustain its great body size it needs to drink 80–200 litres of water each day and needs even more for its baths.

SIZE: 2.4–2.7 metres tall to the shoulder

DIET: Leaves, tree bark, some grasses

FOUND IN: Isolated grasslands and forests throughout Asia

DUGONG
(Dugong dugon)

The dugong is one of the last representatives of a once-thriving group of mammals called sirenians. With each single breath, this strictly herbivorous mammal moves through meadows of seagrass, plunging its head downwards to graze – a bit like an underwater cow.

The dugong's large size protects it from most predators, although crocodiles, orca and some sharks have been known to target its young. Dugongs can grow to be very old indeed, perhaps regularly reaching age 73 or more.

SIZE: Up to 3m long

DIET: Seagrasses

FOUND IN: Indonesia and western regions of the Pacific Ocean

AFRICAN BUSH ELEPHANT
(Loxodonta africana)

Weighing as much as 10 tonnes, the African bush elephant is a giant among land mammals. Females roam the savannah in herds guided by a matriarch (female leader) – often a grandmother with a good memory of where and when the best foods can be found.

An elephant's long incisor teeth (called tusks) are used for digging into streams or moving apart tree trunks in search of food. Its trunk helps it to carefully manipulate things, including stripping leaves from branches to eat. The trunk is a long extension of the top lip and nose and is highly mobile: 40,000 muscles help the trunk to move.

SIZE: Up to 4m tall

DIET: Leaves, tree bark, some grasses

FOUND IN: Plains and grasslands of sub-Saharan Africa

INSECT RAIDERS

Many mammals are specialist hunters of insects, relying on sharp claws and a long tongue to get at and devour their prey. To protect them from angry insect bites and predatory mammals and snakes, many of these insect-hunters have protective spines made from hair. Some, like armadillos and pangolins, even have armour plating.

PINK FAIRY ARMADILLO
(Chlamyphorus truncatus)

The pink fairy armadillo is the world's most secretive armadillo. It is rarely seen by humans. Like an underground torpedo, it burrows through Argentina's dry deserts searching for ants and their larvae.

Like all armadillos, it has plates of bone (called osteoderms) along its back. To help cool down, it pumps blood into these plates, which helps heat escape the body. This is what gives it its distinctive blush colour.

SIZE: 9–11.5cm long

DIET: Ants and insect larvae

FOUND IN: Shrublands of Argentina

GIANT ANTEATER
(Myrmecophaga tridactyla)

Unlike other anteaters, the giant anteater spends most of its life on the forest floor. When it comes across an ant or termite nest, it smashes the nest apart with its sharp claws before lapping up its prey with a long, sticky tongue. This specialist insect-hunter can move its tongue in and out of its mouth 160 times a minute.

The giant anteater can thrive both in cold and warm environments. On particularly cold nights, it can use its tail rather like a blanket.

SIZE: 1.8–2.2m long

DIET: Ants and termites

FOUND IN: Grasslands and rainforests throughout many parts of South America

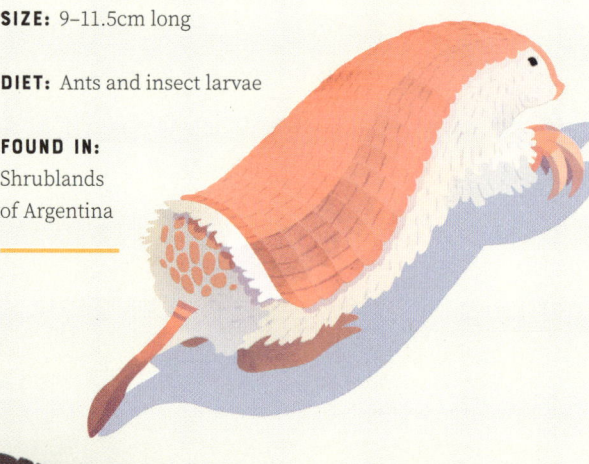

EUROPEAN HEDGEHOG
(Erinaceus europaeus)

The European hedgehog is a firm favourite of gardeners because of its habit of eating garden pests, including slugs, beetles and caterpillars. It seeks this prey out using well-developed ears and an impressive nose. Its sharp jaws deliver a powerful bite.

Like other hedgehogs, the European hedgehog can roll up into a spiky ball when approached by predators. Its pointy spikes are actually modified hairs. Badgers are among only a few animals that can unroll a hedgehog.

SIZE: 20–30cm long

DIET: Mostly insects and slugs

FOUND IN: A wide variety of habitats including woodlands, meadows and gardens across Europe

GROUND PANGOLIN
(Smutsia temminckii)

The overlapping protective scales of the ground pangolin make this the most reptilian-looking mammal in the world. Made of modified hair, its scales afford it protection from a host of predators when rolled up into a ball. The scales on its tail can even be used as makeshift blades to injure would-be marauders.

A thriving illegal trend for using pangolins in alternative medicines has seen many thousands of these animals killed by humans. Like all pangolins, this species is currently drifting toward extinction.

SIZE: 90–110cm long including tail

DIET: Termites and ants

FOUND IN: Savannah woodlands of Southern, Central and East Africa

EVEN-TOED HOOFED MAMMALS

Ungulates are hoofed mammals, and the even-toed ungulates are the group whose hooves are made up of two of their toes, the third and fourth toes. In all, this group contains more than 200 mammals, including many livestock animals such as sheep, goats and cows as well as well-known grassland animals such as many species of deer and antelope and the giraffe and hippopotamus.

COMMON HIPPOPOTAMUS
(Hippopotamus amphibius)

Weighing more than 1.5 tonnes, the common hippopotamus is a superb sinker. On a single breath it dives down, sinks underwater and trots along the lake bottom, pulling up aquatic vegetation to eat with its powerful jaws.

The hippo is extremely territorial. It uses its 40-centimetre-long canine and incisor teeth to scare off rivals and regularly snaps its powerful jaw muscles open and shut in display. In fact, a hippo can bite with twice the force of a lion. This makes it one of Africa's most-feared animals.

SIZE: 1.5m tall to the shoulder

DIET: Grasses

FOUND IN: Rivers, lakes and mangrove swamps across central and southern Africa

NORTHERN GIRAFFE
(Giraffa camelopardalis)

A giraffe's long neck is as much about fighting as it is about feeding. As well as reaching the tallest branches to feed, male giraffes also engage in long bouts of neck-wrestling. They have little hornlike structures called ossicones that can inflict serious wounds. Often the winner is the male that is best at endurance.
The northern giraffe's heart is extra-muscular to pump blood all the way up its neck to the brain. The heart alone weighs 11 kilograms and beats 150 times a minute.

SIZE: 4.3–5.7m tall

DIET: Twigs, leaves, shrubs and some grasses

FOUND IN: Savannah and open woodlands across many parts of Africa

ODD-TOED HOOFED MAMMALS

The odd-toed ungulates have hooves that each consist of a single enlarged middle toe. As well as horses and their relatives, this group includes some of the most threatened animals on Earth, such as the rhinoceros of Africa and the tapirs of South America. Only 17 odd-toed ungulate species survive today.

PLAINS ZEBRA
(Equus quagga)

Each year on the plains of Africa, thousands of battles for social status take place between plains zebras eager to take charge of family groups called harems. Its highly social nature makes the zebra a very intelligent and communicative animal.

No one is quite sure why the zebra has its distinctive stripes. When herds are viewed from afar, the stripes may confuse predators, or it may be that they help zebras communicate with one another. Increasingly, scientists think zebra stripes may be an adaptation to deter biting flies.

SIZE: 1.1–1.45m tall to the shoulder

DIET: Grass, herbs and shrubs

FOUND IN: Grassy plains across eastern and southern Africa

SOUTH AMERICAN TAPIR
(Tapirus terrestris)

The South American tapir is the Amazon's largest surviving native mammal. This secretive animal moves quickly through dense undergrowth and regularly swims and dives to find food. Scientists are currently trying to learn more about its habits using special cameras.

Though large and heavy, the South American tapir is no match for a number of Amazonian predators including jaguars, caiman crocodiles and the green anaconda. Poaching by humans is another threat, meaning that this species is threatened with extinction.

SIZE: 1.8–2.5m long

DIET: Leaves, buds and shoots

FOUND IN: In patchy populations throughout the Amazon rainforest

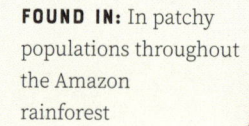

BLACK RHINOCEROS
(Diceros bicornis)

Two horns adorn the dinosaur-like head of the black rhinoceros. These horns are not made of bone but of keratin, the same material found in mammal hair and nails. Sometimes growing more than a metre in length, these impressive weapons can be used for digging, fighting or scaring away predators.

Sadly, human demand for rhino horns has seen the black rhinoceros become one of the world's most threatened creatures. Today, as few as 5,000 may be left in the wild.

SIZE: 140–180cm tall to the shoulder

DIET: Shoots, thorny bushes, fruits and leafy plants

FOUND IN: Isolated grasslands of southern and eastern Africa

BIG CATS

The big cats are part of a captivating group of mammals called carnivorans. All cat species have keen eyesight, sharp claws and sharp teeth that can slot together like scissors to bite through muscle and bone. The big cats are very large and powerful, and can roar.

LION

(Panthera leo)

What solitary carnivorans gain from stealth, the lion gains from teamwork. Family groups, called prides, seek out and stalk a range of mammals, particularly ungulates such as gazelle, zebra, wildebeest, buffalo and giraffes. Scanning vast herds of potential prey with their sharp eyes, lions pick out the weakest individuals and then bring them down with a co-ordinated attack.

Both males and females defend the pride from other nearby lions, but males may engage in longer bouts of energetic fighting. The male's long mane helps him avoid damage to his vulnerable neck and throat.

SIZE: Up to 3m long

DIET: Large mammals, especially ungulates

FOUND IN: Isolated populations in grasslands across sub-Saharan Africa and in western India

TIGER
(Panthera tigris)

The tiger's canine teeth sometimes measure nearly eight centimetres long. They are lined with special nerve endings so that the tiger can use them to feel for its prey's spinal cord before delivering its killer bite.

The camouflage patterns – stripes – on the sides of the tiger's body hint at its secretive way of life. Hidden in the shadows, it creeps slowly and carefully toward unsuspecting prey before leaping almost four metres through the air with its claws outstretched. The prey rarely knows what hit it.

SIZE: Up to 3.3m long

DIET: Large and medium-sized mammals including deer, wild boar and buffalo

FOUND IN: Dense woodlands and forest throughout isolated regions of India, South East and Central Asia

DOGS AND DOGLIKE CARNIVORES

This group of long-snouted carnivores are called canids. The canids can be very social and many species work together in packs to bring down prey. The group includes dogs, foxes, wolves, jackals and more, and there are species of canid found all over the world except Antarctica.

FENNEC FOX
(Vulpes zerda)

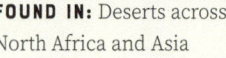

To stop its body from overheating, the fennec fox has huge ears that give off heat. These ears are also sensitive to the sound of insects. If it hears burrowing noises under the surface, it can dig through the sand with powerful claws to catch them.

Sometimes dew forms around the opening of the fennec fox's burrow, which it will eagerly lap up. This dew may be the only water it ever drinks. It gets the rest of the water it needs from its food.

SIZE: Up to 65cm long

FOUND IN: Deserts across North Africa and Asia

DIET: Insects, fruits, birds' eggs and rodents

COYOTE
(Canis latrans)

The coyote is a clever animal that can alter its behaviour depending on its circumstances. Some hunt in family groups, like wolves. Others choose a life alone, hunting more like foxes.

The coyote has adapted well to living alongside humans, so much so that some people consider it a pest – not least because of the disturbing noises it can make at night. The coyote is one of the most vocal of all canids. In all, scientists have named 11 calls including woofs, growls, howls and yelps.

SIZE: 140–175cm long from nose to tail

DIET: Rabbits, rodents, birds, amphibians, lizards and snakes

FOUND IN: Grasslands and, increasingly, urban areas, throughout North and Central America

GREY WOLF
(Canis lupus)

Found throughout the northern hemisphere, the grey wolf is the world's most successful canid. Its secret is in its highly social nature. Grey wolves in a pack can work together to bring down prey, taking it in turns to chase their targets to the point of exhaustion before making the killer blow.

At some point in the last 50,000 years, early humans and wolves learned to co-exist. This partnership would have a great effect on both species. The domestic dogs we share our lives with today are all descended from the grey wolf.

SIZE: 134–210cm long from nose to tail

DIET: Mostly medium and large hoofed mammals

FOUND IN: A wide range of habitats from deserts and grasslands to forests and Arctic tundras, throughout the northern hemisphere

BEARS

Bears have a large, stocky body and a long snout. Though they possess sharp teeth and strong jaws, most species are omnivorous and there are even some herbivores, such as the giant panda, in the group.

GIANT PANDA
(Ailuropoda melanoleuca)

The giant panda is a meat-eater that has turned vegetarian. The powerful jaws used to kill animals many millions of years ago are now used to bite through its preferred diet of bamboo instead. It has a sixth finger (made of a stretched wrist bone) to allow it to grasp bamboo stems while it eats.

Scientists think the giant panda's black-and-white colour helps it camouflage in snowy forests. The distinctive eyepatches may help individual pandas recognise one another from a distance.

SIZE: 1.2–1.9m long

FOUND IN: Mountain ranges of central China

DIET: Bamboo stems

SUN BEAR
(Helarctos malayanus)

Using its giant canine teeth, the sun bear pulls apart bark in search of insect grubs and honey. It laps them up with a very muscular tongue that can measure 25 centimetres long. Its long, sickle-shaped claws are useful for tearing apart termite mounds. It is a small bear that spends much of its time in the treetops. Each paw has bare patches that help it with grip whilst climbing.

Unlike most other bears, the sun bear has predators. Tigers, leopards and even certain pythons have been known to eat this species.

SIZE: 1.2–1.5m long

FOUND IN: Isolated tropical forest habitats of Southeast Asia

DIET: Mostly fruits, insect larvae and honey

POLAR BEAR
(Ursus maritimus)

Alone, the polar bear walks the frozen Arctic. With each breath it scans for the distant smell of seals, which it can detect from more than 1.6 kilometres away. Its eyes scan the horizon for movement. Its ears listen for telltale noises made by prey. It is the planet's most icy killer.

To keep from freezing, the polar bear has dense fur and a second layer rather like a woolly jumper. The hairs in this outer layer look white but they are actually see-through.

SIZE: 1.8–3m long from nose to tail

DIET: Seals

FOUND IN: Within the Arctic Circle and nearby land masses including Newfoundland, Canada

DIVERSE CARNIVORES

We have already met many of the group of mammals called carnivorans on these pages, including the seals and their relatives, the cats, the dogs and their relatives, and bears, but there is a broad range of other animals in this group that has taken to the role of killing and eating other creatures.

MEERKAT
(Suricata suricatta)

Watchful meerkats stand tall searching the horizon for predators while other members of their mob explore the surroundings for invertebrate prey. If an attacker is spotted, these sentry meerkats let out a special bark or whistle which sends the whole mob scattering for cover.

The meerkat's social nature extends to rearing babies. Adult meerkats regularly teach their young the way to eat scorpions without getting stung and individuals will often share childcare.

SIZE: 60–75cm long from nose to tail

FOUND IN: Deserts of southern Africa

DIET: Insects, spiders and scorpions as well as small snakes and lizards

SPOTTED HYENA
(Crocuta crocuta)

The spotted hyena can work as part of a pack to chase animals over long distances. By relay-running with other group members, it can keep up speeds of over 60 kilometres per hour over many kilometres. This is enough to exhaust even the healthiest wildebeest.

But the spotted hyena is more than just a killer. It has powerful stomach acids which mean it can occasionally scavenge rotting animals without becoming poisoned. In fact, it regularly digests bones and the tough skin of animal hides.

SIZE: 95–136cm long from nose to tail

FOUND IN: Throughout sub-Saharan Africa

DIET: Large mammals including wildebeest, gazelle and buffalo

GIANT OTTER
(Pteronura brasiliensis)

With webbed feet, dense fur and a powerful tail, the giant otter is one of the world's best-adapted fish-eaters. Its long whiskers are used to detect the movement of fish, which it snaps at with catlike jaws. Sometimes it will take bigger prey, including snakes and small caimans.

As well as being an expert hunter, the giant otter is highly social. Individuals in a family group will sleep, play and travel together to hunt.

SIZE: 1–1.7m long from nose to tail

FOUND IN: South America, particularly in and along the Amazon river

DIET: Fish, crustaceans and small caiman

INSECT-EATERS

The insectivores, or insect-eaters, are a group of mammals that all have strong and sharp teeth to kill prey, and live highly active lifestyles. They have taken the role of insect-eater to impressive levels through a host of unusual adaptations, but they do eat some other prey too.

HISPANIOLAN SOLENODON
(Solenodon paradoxus)

The Hispaniolan solenodon has a bone running through the middle of its snout that makes it extra firm. This means that it can push the snout through tough soil without getting injured.

When a creature is seized in its jaws, the Hispaniolan solenodon does something almost totally unique among mammals. It can deliver a venomous bite. Just half a milligram of its venom is enough to kill a mouse in minutes.

SIZE: 49–72cm long from nose to tail

DIET: Mostly insects and spiders, as well as mice, lizards and worms

FOUND IN: Undisturbed moist forests on the Caribbean island of Hispaniola

ETRUSCAN SHREW
(Suncus etruscus)

Weighing the same as a single playing card and measuring not much longer than a paper clip, the Etruscan shrew is arguably the world's smallest mammal. It dives in and out of the undergrowth, pouncing upon insects which it devours speedily.

Smaller mammals have to burn up more energy to stay alive than bigger mammals. So its tiny body size means that the Etruscan shrew must eat all day long. If it fails to eat twice its body weight in food each day it will die of starvation.

SIZE: 4cm long not including tail

DIET: Insects

FOUND IN: Deciduous forests and grasslands throughout Europe and Asia

EUROPEAN MOLE
(Talpa europaea)

The European mole quite literally swims through soil. Each of its well-muscled front legs ends with a wide spadelike hand with sharp claws, and its body is streamlined like a torpedo. Each mole digs a network of tunnels 200 metres long which it patrols for invertebrates, particularly worms. This is equivalent to a human digging for 2.5 kilometres using only their bare hands.

Once the mole has seized a worm in its jaws, it delivers a sharp bite into which drips saliva that is toxic to worms. This poison stops the worm from escaping. The mole then carries its incapacitated prey to an underground chamber where its food is stored.

Like other burrowing mammals, the European mole has very small eyes. Each eyeball only measures about one millimetre across, which is not much bigger than the head of a pin.

SIZE: 11–16cm long from nose to tail

DIET: Mostly earthworms, centipedes and insects

FOUND IN: Soft soils throughout Europe

MARSUPIALS

Marsupials are a group of mammals that have special pouches in which they keep their young. Marsupials once thrived throughout the southern hemisphere but today most of these fascinating mammals live in Australia.

KOALA
(Phascolarctos cinereus)

The koala gets so little nutrition from its diet of eucalyptus leaves that it spends up to 20 hours each day sleeping to save energy. It holds onto branches with specially adapted grasping toes and each paw has two thumbs to provide it with extra grip.

During the breeding season, the koala communicates with a loud bellowing noise. This low-frequency sound travels impressively well through dense forests, meaning that koalas can hear potential mates or rivals approaching from far away.

SIZE: 60–85cm long

DIET: Mostly eucalyptus leaves

FOUND IN: Eucalyptus forests throughout eastern and southeastern Australia

RED KANGAROO
(Osphranter rufus)

In a single leap, the red kangaroo can spring almost 10 metres through the air. It achieves this impressive feat by recycling the energy from the impact of its previous jump, using elastic tendons in the legs. Each time it lands, these tendons catapult the kangaroo back into the air.

Like all marsupials, the kangaroo has very strong forearms. These are crucial in the first days of life, when the tiny baby (called a joey) has to climb and scramble up through its mother's fur to reach the safety of her pouch.

SIZE: 2.5–2.8m long from nose to tail

DIET: Fresh grasses and flowering plants

FOUND IN: Scrublands, grasslands and deserts throughout central and western Australia

LONG-NOSED BANDICOOT
(Perameles nasuta)

A female long-nosed bandicoot is pregnant for just 12 days before giving birth to tiny babies that scurry into her pouch. The young stay here for another 50 days before they are weaned and must seek a life of their own, hunting insects and fungi on the forest floor.

The long-nosed bandicoot has a very long snout which it plunges into soil to detect buried insects. This behaviour leaves a cone-shaped pocket in the ground that helps scientists know where these secretive animals have been.

SIZE: Up to 40cm long from nose to tail

DIET: Insects, plant roots and fungi

FOUND IN: Rainforests and grasslands across eastern Australia

COMMON WOMBAT
(Vombatus ursinus)

Unlike most marsupial mammals, the common wombat has a pouch that opens backwards. This stops soil being kicked into its pouch whilst it digs through soil to make its underground network of tunnels. The common wombat digs using a mixture of sharp claws and piercing rodent-like teeth.

When threatened by predators, the common wombat flees back to its tunnels and blocks the entrance with its armoured bottom. Its rear-end is shielded with a layer of cartilage, the same soft bone found in human noses.

SIZE: Approx. 1m in length

DIET: Grasses, bushes, herbs and tree bark

FOUND IN: Common in a wide range of habitats throughout south-eastern Australia

EGG-LAYERS

One tiny group of mammals lays eggs. These are the so-called 'monotremes'. This dwindling part of the mammal family tree gives scientists a unique understanding of how early mammals evolved from reptile-like mammals more than 300 million years ago.

SHORT-BEAKED ECHIDNA
(Tachyglossus aculeatus)

The short-beaked echidna has a tongue that can stretch an amazing 18 centimetres from out of its snout and can be moved around a bit like a long finger. This tongue is covered in a gluelike substance to trap unwary insects. Sharp teeth inside the mouth and throat trap insects as the echidna draws its tongue back in.

Unlike other mammals, the short-beaked echidna does not use stomach acids to digest prey. Instead, special spines on its tongue and the roof of its mouth help grind up food into tiny, digestible chunks.

SIZE: 30–45cm long

DIET: Ants and termites

FOUND IN: Throughout Australia and parts of New Guinea

PLATYPUS
(Ornithorhynchus anatinus)

The platypus is so strange that, when scientists first described it more than 200 years ago, many people thought it was a hoax. This was, after all, an egg-laying mammal, with a bill like a bird, that 'sweated' milk for its babies to drink. Only later would scientists realise that monotremes are survivors from an early age of mammals that lived alongside early dinosaurs – a time when many mammals were far more reptile-like in their nature.

Today, the platypus has made a freshwater life for itself which few modern mammals can rival. It is one of only a handful of mammals able to detect tiny amounts of electricity generated by the movement of fleeing prey while underwater. It is so sensitive to these electrical fields that it needs no other senses as it swims. It keeps its eyes, ears and nose closed while underwater.

SIZE: 43–50cm long from bill to tail

DIET: Aquatic invertebrates including worms, insect larvae and crayfish

FOUND IN: Small streams and rivers across isolated regions of eastern Australia

INDEX

A

ALBATROSS, WANDERING 119
ALLIGATOR
    AMERICAN 92
    CHINESE 92
ANACONDA, GREEN 94, 172
ANEMONE 71
    SNAKELOCKS 12
ANGELFISH, EMPEROR 70
ANOLE, BROWN 99
ANT 7, 32, 125, 168
    ARGENTINE 33
ANTEATER, GIANT 168
ANTELOPE 170
APE 146, 152-53
ARACHNID 7, 30
ARMADILLO 90
    PINK FAIRY 168
ARTHROPOD 24, 28, 30, 32, 44, 45
AXOLOTL 82, 83
AYE-AYE 148, 149

B

BADGER 169
BANDICOOT, LONG-NOSED 185
BARNACLE, POLI'S STELLATE 27
BARRACUDA, GREAT 68
BAT 38, 44, 154-55
    BUMBLEBEE 142
    COMMON VAMPIRE 154
    FRUIT 50
    LESSER HORSESHOE 155
    SPOTTED 155
BEAR 178-79
    POLAR 142, 179
    SUN 178
BEAVER, EURASIAN 160
BEE 6, 7
    WESTERN HONEY 32
BEETLE 6, 7, 34-35, 65, 169
    ROYAL GOLIATH 34

BICHIR 58
    GUINEAN 58
BIRD OF PARADISE 106
    WILSON'S 133
BIRD OF PREY 136-39, 163
BIVALVE 18, 19
BONEFISH 60
BONOBO 152
BOWFIN 59
BRACHIOPOD 18, 19
BRITTLESTAR, COMMON 17
BULLFROG, AFRICAN 78
BUTTERFLY 7, 38-39, 150
    MONARCH 39

C

CADDIS FLY, LAND 38
CAECILIAN 76, 81
    TAITA AFRICAN 81
CAIMAN 63, 94, 172, 181
    SPECTACLED 93
CANARY, ATLANTIC 129
CANID 176, 177
CAPUCHIN, CRESTED 151
CAPYBARA 160
CARP, GRASS 62
CASSOWARY, SOUTHERN 109
CAT, BIG 174-75
CATFISH, BLACK BULLHEAD 62
CENTIPEDE, AMAZONIAN GIANT 44
CEPHALOPOD 22-23
CHAMELEON, JACKSON'S 98
CHAN'S MEGASTICK 36
CHIMAERA 48, 56
CHINCHILLA, SHORT-TAILED 158
CLAM 6, 56, 88
    GIANT 18
    PACIFIC RAZOR 18
CLOWNFISH, COMMON 71
COBRA, KING 95
COCKATOO, SULPHUR-CRESTED 134
COLOBUS, KING 144
CONDOR, ANDEAN 136
CORAL 14, 70
    GROOVED BRAIN 14
    OUTCROP 53
    REEF 20, 23, 25, 48, 86, 88, 95
    USED AS ANCHOR 16

COW 154, 170
COYOTE 176
CRAB 16, 50, 54, 55, 56, 88
CRAB, HALLOWEEN 24
CRICKET, EUROPEAN MOLE 36
CROCODILE 58, 86, 87, 92, 93, 166
    SEE ALSO CROCODILIAN AND CAIMAN
    SALTWATER 86, 93
CROCODILIAN 92-93
    SEE ALSO CROCODILE
CROW 106, 137
    NEW CALEDONIAN 130
CRUSTACEAN 6, 7, 24-27, 60, 65, 72, 93, 117
    SEE ALSO KRILL
CUCKOO, COMMON 113
CUTTLEFISH 22
    COMMON 23

D

DAMSELFLY 42
DEER 92, 94, 102, 170
DEMOISELLE, BANDED 42
DOG 176, 177
DOLPHIN 63, 156, 164
DOVE 112, 139
    ROCK 112
    SPOTTED 112
DRAGONFLY 42
DUCK 114-15, 127
    MANDARIN 115
DUGONG 166

E

EAGLE 58, 146
    PHILIPPINE 138
EARTHWORM 81
    GIANT GIPPSLAND 8
EASTERN BUZZARD 137
ECHIDNA 142
    SHORT-BEAKED 186
ECHINODERM 16
EEL 48, 60, 64
    GIANT MORAY 61
EGRET, GREAT WHITE 116
ELEPHANT 143, 166-67
    AFRICAN BUSH 167
    ASIAN 166

EMU 108

F

FALCON, PEREGRINE 139
FALSE MOORISH IDOL 71
FIREFLY, BIG DIPPER 35
FLAMINGO, GREATER 117
FLATWORM 10
    PERSIAN CARPET 11
FLEA 65
    SNOW 45
    WATER 65
FLY 6, 40-41
FLY, VINEGAR 40
FLYING DRAGON, COMMON 100
FLYING FISH, ORNAMENTED 48, 69
FLYING FOX, LARGE 154
FOX, FENNEC 176
FROG 20, 28, 76, 77, 78-79, 116, 137
    DYEING DART 79
    NORTHERN GLASS 79
    POISON ARROW 77, 79
    RED-EYED TREE 78

G

GALAGO, BROWN GREATER 146
GAME BIRD 110-11
GAR 59
    ALLIGATOR 59
GASTROPOD 20-21
GECKO 86
    TOKAY 98
GHARIAL 92
GHOST SHARK 56
GILA MONSTER 87, 102
GIRAFFE 170, 174
    NORTHERN 171
GOAT 122, 170
GOLIATH BIRDEATER 28
GOOSE 114-15, 127
    BAR-HEADED 115
GORILLA, WESTERN 152, 153
GOSHAWK, NORTHERN 139

H

HAGFISH, ATLANTIC 57
HAMMERHEAD, GREAT 52
HARE 127, 162
    SNOWSHOE 162
HARVESTMAN, GIANT LAOTIAN 31
HEDGEHOG, EUROPEAN 169
HELLBENDER 83
HIPPOPOTAMUS, COMMON 170
HORSE 50, 154, 172
HOVERFLY, AMERICAN 40
HUMMINGBIRD, BEE 123
HYENA, SPOTTED 181
HYMENOPTERA 32-33

I

IGUANA, GREEN 101

J

JACKRABBIT, BLACK-TAILED 163
JAY, CALIFORNIA SCRUB 131
JELLYFISH 6, 10, 12-13, 89
JERBOA, LONG-EARED 158
JUNGLEFOWL, RED 110

K

KANGAROO, RED 184
KINGFISHER, COMMON 124
KOALA 184
KOMODO DRAGON 87, 102
KRILL 55, 121, 157, 165

L

LADYBIRD, HARLEQUIN 35
LAMP SHELL 18, 19
LAMPREY, EUROPEAN RIVER 57
LANGUR, SACRED 145
LANTERNSHARK, DWARF 53
LEECH, MEDICINAL 8
LEMUR 148-49
    RING-TAILED 148
LION 142, 174

LIONFISH 66
    RED 67
LIZARD 86, 87, 98-103, 109
    DRACO 86
    MEXICAN MOLE 103
    SHINGLEBACK 101
LOBSTER 54
    AMERICAN 24
LORIKEET, RAINBOW 134
LUNGFISH 48
    AUSTRALIAN 73

M

MACAQUE, JAPANESE 145
MACAW, HYACINTH 135
MAGPIE, EURASIAN 130
MAKO, SHORTFIN 50
MALLARD 114
MAMBA, BLACK 96
MANDRILL 144
MANTIS SHRIMP, PEACOCK 25
MANTIS, WALKING FLOWER 37
MARLIN, BLACK 68
MARMOSET, WESTERN PYGMY 150
MARSUPIAL 142, 143, 184-85
MATA MATA 90
MEERKAT 180
MILLIPEDE 44
MOLE 127
    EUROPEAN 183
MOLE-RAT, NAKED 158, 159
MOLLUSC 6, 18, 20, 22, 54, 164
MONKEY 138, 142, 144-47, 150-51
    SEE ALSO APE
    BROWN HOWLER 150
MONOTREME 142, 143, 187
MOSQUITO 7
    YELLOW FEVER 43
MOTH 7, 38-39
    GROTE'S TIGER 38
MOUSE 127, 182
    AFRICAN PYGMY 159
MYRIAPOD 44

N

NAUTILUS	22
CHAMBERED	23
NEWT	76, 77
GREAT CRESTED	82
NIGHTINGALE	129
NIGHTJAR, EUROPEAN	122

O

OCTOPUS	7
GREATER BLUE-RINGED	22
ORANGUTAN, SUMATRAN	152
ORCA	156, 166
OSPREY	137
OSTRICH	108
OTTER	131
GIANT	181
OWL	125, **126-27**, 146, 158
BARN	126
BLAKISTON'S FISH	126
BURROWING	127
LITTLE	158
SNOWY	127

P

PADDLEFISH	72
AMERICAN	72
PANDA, GIANT	178
PANGOLIN	168
GROUND	169
PARROT	79, **134-35**
PARROTFISH, BICOLOUR	70
PEAFOWL, INDIAN	111
PELICAN, PERUVIAN	119
PENGUIN	**120-21**, 165
AFRICAN	120
EMPEROR	106, 120
GENTOO	121
LITTLE	121
PIGEON	20, 112, 139
PIKA, AMERICAN	162
PIKE, NORTHERN	64
PIRANHA, RED-BELLIED	63
PLATYPUS	142, 187
PORTUGUESE MAN-OF-WAR	13
PUFFIN, ATLANTIC	118

Q

QUAIL, KING	111

R

RABBIT	162
SUMATRAN STRIPED	163
RAT	20, 50
BROWN	161
RATFISH, SPOTTED	56
RATTLESNAKE, TIMBER	94
RAVEN	106
COMMON	131
RAY	19, 48, **54-55**, 60
BLUESPOTTED RIBBONTAIL	54
GIANT OCEANIC MANTA	55
REEDFISH	58
REMIPEDE, GODZILLA	26
RHEA, AMERICAN	109
RHINOCEROS	172
BLACK	173
RIFLEBIRD, MAGNIFICENT	132
ROBBERFLY, VIOLET BLACK-LEGGED	41
ROCK AGAMA, MWANZA FLAT-HEADED	99
RODENT	44, 137, 143, **158-61**
ROTIFER	6, 15

S

SALAMANDER	28, 76, **82-83**
FIRE	82
SALMON, ATLANTIC	64
SCALLOP, BAY	19
SCORPION	6, 30, 180
EMPEROR	30
TANZANIAN TAILLESS WHIP	30
SCORPIONFLY	40
COMMON	41
SEA COW	166
SEA CUCUMBER, SNAKE	17
SEA GOOSEBERRY	13
SEA LION, CALIFORNIA	164
SEA SNAKE	50, 86
OLIVE	95
SEA STAR, SUNFLOWER	16
SEA WASP	12
SEABIRD	**118-19**
SEAHORSE, DWARF	66
SEAL	72, **164-65**, 179
ANTARCTIC FUR	165
SOUTHERN ELEPHANT	165
SHARK	48, **50-53**, 68
AS PREDATOR	61, 88, 95, 166
AS PREY	93
FILTER-FEEDING	60
GREAT WHITE	48, 49, 51, 52
POPULATION	19
TIGER	50, 52, 88
WHALE	52
SHEEP	170
SHELLFISH	**18-19**, 25
SHREW, ETRUSCAN	142, 182
SHRIMP	55, 64
PACIFIC GIGANTIC SEED	27
TADPOLE	26
SILVERFISH	45
SLOW LORIS, BENGAL	147
SLUG	7, 20, 169
LEOPARD	20
SMELT, POND	65
SNAIL	6, 7, 20, 88, 93, 114
AFRICAN GIANT	21
SNAKE	87, **94-97**
SEE ALSO SEA SNAKE	
AS PREDATOR	146, 158, 168
AS PREY	28, 78, 90, 109, 116, 137, 181
FLOWERPOT	97
MALAGASY LEAF-NOSED	97
SOLENODON, HISPANIOLAN	182
SONGBIRD	**128-29**
SPANISH DANCER	20
SPHAERODACTYLUS ARIASAE	86
SPIDER	6, 7, **28-29**, 30, 41, 44
SEE ALSO ARACHNID	
CAROLINA WOLF	29
DARWIN'S BARK	28
PEACOCK JUMPING	29
SPIDER MONKEY, BLACK-HEADED	151
SPONGE	14, 20, 70
BATH	14
SPOOKFISH, PACIFIC	56
SPRINGTAIL	45

SQUID	50, 69, 142, 165	LOGGERHEAD	88
GIANT	22, 156	YANGTZE GIANT SOFTSHELL	91
SQUIRREL			
EASTERN GREY	161		
GROUND	127		
STARFISH		UNGULATE	170-73, 174
SEE SEA STAR, SUNFLOWER AND BRITTLESTAR, COMMON		URCHIN, FLOWER	16
STARLING, COMMON	128		
STINGRAY	52, 54, 60		
STORK, PAINTED	116	VIPER, PIT	86, 94
STURGEON	72	VULTURE, RED-HEADED	136
BELUGA	72		
SWAN	114		
BLACK	114		
		WALRUS	164
		WASP	7
		EMERALD COCKROACH	32
TADPOLE	76, 77, 80	WATER BIRD	63, 90
TAIPAN, INLAND	96	WHALE	48, 51, 60, 142, 156-57
TAPIR	94, 172	BLUE	142, 157
SOUTH AMERICAN	172	SPERM	142, 156
TARANTULA	28	WOBBEGONG, SPOTTED	53
TARDIGRADE	6, 15	WOLF	131, 176
TARPON, ATLANTIC	60	GREY	177
TARSIER, PHILIPPINE	146	WOMBAT, COMMON	185
TERMITE	81, 152	WOODLOUSE, DESERT	25
MOUND	30, 178	WOODPECKER	124, 148
NEST	168	PILEATED	125
THORNY DEVIL	100	WORM	6, 8-11, 81, 90, 137, 183
TIGER	111, 142, 175, 178	ACORN	10, 11
TOAD	76, 77, 80	ARROW	10
COMMON	80	BOBBIT	8, 9
GOLDEN	80	PINK VELVET	10
TORPEDO, ATLANTIC	54	WREN, CACTUS	128
TORTOISE	90		
GOPHER	91		
TOUCAN	124, 135		
TOCO	124	ZEBRA	174
TROUT, RAINBOW	65	PLAINS	172
TUATARA	86, 103	ZOOPLANKTON	72
TUNA	48, 49		
PACIFIC BLUEFIN	69		
TURKEY, WILD	110		
TURTLE	86, 88-91		
ALLIGATOR SNAPPING	90		
GREEN SEA	88		
LEATHERBACK	87, 89		

Encyclopedia of Animals © 2019 Quarto Publishing plc.
Text © 2019 Jules Howard.
Illustrations © 2019 Jarom Vogel.

First published in 2019 by Wide Eyed Editions, an imprint of The Quarto Group.
This paperback edition first published in 2024 by Wide Eyed Editions, an imprint of The Quarto Group.
1 Triptych Place, London, SE1 9SH, United Kingdom.
T +44 (0)20 7700 9000 www.Quarto.com

The right of Jules Howard to be identified as the author of this work and Jarom Vogel to be identified as the illustrator of this work has been asserted by them in accordance with the Copyright, Designs and Patents Act, 1988 (UK).

All rights reserved.

No part of this publication may be reproduced, stored in a retrieval system or transmitted, in any form, or by any means, electrical, mechanical, photocopying, recording or otherwise without the prior written permission of the publisher or a licence permitting restricted copying.

A catalogue copy of this book is available from the British Library.

ISBN 978-0-7112-9159-1
eISBN 978-1-78603-461-8

The illustrations were created digitally.
Set in Bourton and Source Serif Pro.

Published by Rachel Williams and Jenny Broom
Designed by Nicola Price
Project edited by Katy Flint
Copy-edited by Catherine Brereton
Production by Dawn Cameron

Manufactured in Guangdong, China TT102024

9 8 7 6 5 4 3